The Silent Chauffeur

The Silent Chauffeur

How Self-Driving Cars Will Rewire the Rules of Cities, Power, and Personal Agency

Rowan Hale

Mindful Pages

Published in 2025

ISBN: 9789371776684 (PB)
ISBN: 9789371772587 (eBook)

Published by

Mindful Pages
Imprint of Alpha Editions LLC
312 W. 2nd St #1834
Casper, WY 82601, USA

This book was developed using advanced writing tools to support the drafting and ideation process. All content has been thoughtfully curated, refined, and reviewed to ensure quality and relevance.

Content

Introduction
The Machine at the Crossroads

There is something deeply disquieting about observing a car move without a driver. It navigates intersections, adapts to curves, and brakes just shy of a crosswalk—all seemingly without guidance. The wheel turns, yet no hand is visible; the machine obeys, but to whose command? The choreography is flawless, the scene eerily quiet, and the implications quietly seismic.

This book begins not with the engineering feat itself, but with a question: what does it mean when an act as intimate and human as movement is ceded to a machine? For over a century, the automobile has symbolised personal freedom, industrial ingenuity, and cultural identity. We name our cars, customise them, bond with them. Driving has been ritual, work, and passage. Now, that ritual is quietly slipping away. The silent chauffeur—a complex system of sensors, algorithms, and corporate logic—has assumed control.

Engineering marvels notwithstanding, this transition is more than a technical story. Autonomous vehicles (AVs) do not merely replace drivers; they reconfigure control, reshape cities, and alter human behaviour. They are, quite literally, instruments of power. This book is less concerned with how self-driving cars operate and more with what they do: to laws, labour, urban life, and, most importantly, to ourselves. As machines take on decision-making in motion, we must ask: who sets their priorities? What is lost when friction disappears, and what dependencies emerge in its absence?

We stand at a philosophical inflection point. For the first time since the invention of the internal combustion engine, the act of moving through the world is delegated to an external intelligence. Where once the car amplified human agency, the AV risks diminishing it. Safety and efficiency come at the cost of judgement; optimisation may suppress the unpredictability that is distinctly human.

Yet the rise of AVs is neither inevitable nor neutral. It is shaped by choices in code, policy, design, and capital—a future quietly authored by engineers, corporations, and governments. Public discourse remains shallow, preoccupied with liability and lane-changing, while the deeper questions go unasked: how will autonomy be governed? Whose values will it encode? How do humans remain present in systems designed to operate without them?

Urban design is already shifting in response. Roads are remodelled for precision, not pedestrians. Unpredictability is noise; curbs and drop-offs are redesigned for seamless flow. Crosswalks become optional, eye contact irrelevant. Spaces once dedicated to spontaneity—plazas, parks, street corners—are being supplanted by infrastructures of efficiency. This is more than a change in traffic; it is a subtle reconfiguration of civic life.

Labour, too, is transformed. Driving is work, identity, rhythm. Millions earn a living behind the wheel. When machines succeed, what becomes of that work, of those communities, of the sense of purpose embedded in a daily commute? Displacement is not merely economic; it is existential. Policy often lags, reactive and blind to the long arc of automation's social impact.

Data is the lifeblood of AVs. Every journey produces streams of information: routes, speeds, companions, timings. This data is mined, monetised, sometimes weaponised. Movement becomes metadata; consent perfunctory. Surveillance is no longer stationary—it moves, predicts, and persists. The ephemeral act of travel is transformed into enduring, analysable records.

Ethical questions run just as deep. AVs must make trade-offs and moral judgments. Who is protected in an unavoidable collision? Whose life is statistically prioritised? These decisions are encoded in lines of software, constrained by corporate liability, and fed by imperfect datasets. The logic of the machine becomes, in effect, the morality of the street.

Yet within this tension lies possibility. Approached critically, autonomy could expand mobility for underserved populations, reduce emissions, and improve urban livability. Ownership models could be reimagined to be equitable. Dignity, not just efficiency, could be embedded in code. But such outcomes demand vigilance, imagination, and civic engagement. We must not become passive passengers in our own societies.

Across ten chapters, this book explores the dimensions of this transformation. It begins with *The Illusion of Arrival*, which unpacks the myth of technological inevitability and exposes the careful orchestration behind the so-called self-driving revolution. Subsequent chapters trace the migration of agency from human to algorithm, interrogate the ethics embedded in machine decision-making, examine the urban geometries AVs produce, and analyse impacts on labour, surveillance, and behaviour. Each chapter builds toward a single aim: to reclaim the human lane—not nostalgically, but as a forward-looking framework for meaningful autonomy.

This is not a neutral account. It is written with urgency because decisions made quietly today will shape the world tomorrow. It is written with hope, too, for when technology is subjected to democratic scrutiny and guided by humane values, it can serve inclusion and creativity. But such an outcome is not automatic; it must be fought for, argued for, and designed for.

The silent chauffeur has arrived, gliding through our cities and quietly rewriting the rules of movement. We may choose to remain silent, lulled by convenience, novelty, or promises of safety. Or we can ask: what kind of autonomy do we desire? Who decides it? And how will we maintain presence, voice, and agency in an age of driverless machines?

This book is an invitation to take back the wheel—not literally, but politically, ethically, and imaginatively. For when machines begin to drive, the paths they follow will depend, more than ever, on the humans still watching the road.

Chapter 1
The Illusion of Arrival

For years, the promise of the self-driving car has shimmered on the edge of reality, part science fiction, part imminent breakthrough. From glossy concept videos to keynote speeches at tech expos, autonomous vehicles (AVs) have been positioned as the crown jewel of innovation—a transformative solution to traffic deaths, inefficiency, and urban congestion. The narrative is seductive: a world where machines glide flawlessly through cities, liberating time, reducing accidents, and democratising mobility. Yet beneath the sheen lies a more complicated truth. Far from a sudden leap into the future, the arrival of self-driving cars has been incremental, uneven, and laden with compromise. What appears to be a technological climax is, in fact, a carefully staged illusion—an artefact of storytelling, branding, and selective visibility.

The vehicles have changed, certainly. What once resided in university labs and defence contracts has now taken to public streets. Pilot programmes in Phoenix, San Francisco, and Beijing offer rides without drivers. Sensors bristle from rooftops. Lidar pulses quietly from hidden mounts. The aesthetic of autonomy has entered the urban vernacular. But to assume that this visual presence signals maturity is to mistake theatre for substance. Most AVs today operate under a patchwork of constraints: geofencing, weather limitations, human remote monitoring. Their presence may suggest autonomy, but their behaviour often reflects a delicate choreography of caveats and exceptions. We are not yet witnessing a technological arrival; we are experiencing a managed rollout, one that balances perception with progress.

The illusion persists, in part, because Silicon Valley has mastered the art of myth-making. The self-driving car has been framed not as a product in development, but as an inevitability—an object of faith rather than scrutiny. This framing taps into a deeper cultural narrative: the machine as saviour. It promises to deliver us from the frailties of human error, the chaos of urban traffic, and the dull drudgery of the daily commute. In doing so, it elevates technology beyond mere tools and repositions it as a force of moral correction. The AV is not just convenient; it is good. It is rational. It is the logical endpoint of progress.

Such mythologies are not new. The automobile itself was once heralded as a liberator, a symbol of modern freedom. Its mass adoption in the 20th century reshaped cities, economies, and identities. But it also produced unforeseen consequences: suburban sprawl, pollution, traffic fatalities, social fragmentation. That history should temper our enthusiasm for the next transport revolution. For while the AV is often described as a disruption, it is just as much a continuation—an extension of the same logic that privileges efficiency over intimacy, speed over serendipity, and automation over ambiguity.

To speak of a vehicle as "self-driving" is, at best, imprecise and, at worst, misleading. No car drives itself in a vacuum. Behind every AV lies an invisible lattice of dependencies: GPS satellites, real-time maps, cloud-based analytics, regulatory frameworks, and, crucially, human intervention. Even the most advanced systems rely on remote operators, machine learning engineers, and vast volumes of training data curated by people. Autonomy, in this context, is not independence. It is a redistribution of effort—from driver to developer, from manual to digital labour. The vehicle may lack a steering wheel, but the system is never truly alone.

The term "self-driving" flattens this complexity. It suggests a singular, monolithic capability when in fact autonomy exists on a spectrum—one shaped by context, environment, and contingency. The Society of Automotive Engineers (SAE) defines six levels of driving automation, from driver assistance to full autonomy. But even this taxonomy fails to capture the messiness of real-world implementation. An AV that performs admirably on dry roads in California may falter in the snow-laden alleys of Montreal or the bustling unpredictability of Delhi. Autonomy is not just a technical challenge; it is a cultural, infrastructural, and environmental one.

And yet, the desire to believe in the inevitability of AVs persists—not just among technologists, but among consumers. The culture of convenience has trained us to welcome any service that promises to save time, reduce effort, or simplify complexity. This predisposition is not irrational. After all, many of the technologies that now define modern life—smartphones, GPS, streaming platforms—entered our routines under the same banner. But convenience is not neutral. It reconfigures relationships, often in ways that go unnoticed until they are irreversible.

In the case of AVs, convenience comes at the cost of control, privacy, and civic accountability. To enter a driverless vehicle is to submit to a system: to trust that its software will navigate safely, that its data will be handled ethically, that its decisions will align with human values. But unlike traditional transport modes, where risks and responsibilities are visible and localised, AVs obscure causality. When something goes wrong—when a pedestrian is struck, or a vehicle swerves unexpectedly—pinpointing accountability becomes an exercise in forensic ambiguity. Was it the algorithm? The data set? The human who coded the decision tree? Or the city that licensed the pilot zone?

This opacity has profound implications. It shifts the burden of proof from institutions to individuals, from systems to users. It asks the public to accept risk without clarity, to trust without access. And in doing so, it deepens a broader pattern in the digital age: the decoupling of use from understanding. Much as we swipe and click without grasping the mechanics beneath the screen, we may soon ride without knowing who—or what—is truly in charge.

The challenge, then, is not to reject AVs outright, but to interrogate the terms under which they are being introduced. We must ask why their limitations are underreported, why their pilots are often shrouded in NDAs, why regulatory frameworks lag behind deployment. We must scrutinise the narratives that equate machine control with safety, and automation with progress. And we must resist the framing of scepticism as Luddism. Critical engagement is not opposition. It is the responsibility of citizens living through a shift as momentous as the industrial revolution.

Consider the cities already adapting to accommodate AVs. In parts of Arizona, local authorities have granted testing permissions with limited public oversight. In China, AV development is tightly entwined with state objectives, creating synergies between data capture and governance. In Europe, regulators have been more cautious, demanding transparency and safety benchmarks before permitting large-scale trials. These differences reveal the geopolitical stakes of AV deployment: it is not merely a question of technology readiness, but of governance philosophy, risk tolerance, and civic values.

Moreover, the way AVs interact with existing infrastructure exposes the fragility of their independence. A construction zone without updated maps can render a vehicle helpless. A jaywalking pedestrian can confound its decision tree. A malfunctioning sensor can provoke cascading failures. These are not fringe scenarios; they are everyday realities. And they highlight a central paradox: the more we pursue full automation, the more brittle our systems may become—overfitted to perfect conditions, yet vulnerable to the unpredictability of lived environments.

That unpredictability is not a flaw to be engineered out. It is the essence of urban life. Cities are messy, pluralistic, alive. They thrive on ambiguity, negotiation, and informal cues—the very things machines struggle to interpret. The rise of AVs challenges this ethos. It proposes a city designed not for humans, but for seamless machine operation. A city where movement is frictionless, but perhaps also soulless. Where unpredictability is not accommodated, but eliminated.

This chapter will explore, in detail, how the myth of the AV's arrival has been constructed and sustained. It will trace the path from prototype to pavement, unpack the language used to frame progress, and dissect the systemic dependencies that belie claims of autonomy. It will examine how consumer psychology is leveraged to drive adoption, and how convenience is wielded as both carrot and cover. Above all, it will argue that the story of self-driving cars is not yet written—and that by exposing the illusion, we reclaim the capacity to shape what comes next.

To do so requires clear eyes, historical memory, and intellectual humility. The AV may not have fully arrived, but its consequences are already rippling through law, labour, infrastructure, and imagination. The question is no longer whether we will coexist with driverless systems, but on what terms. Will we participate in their shaping, or will we watch passively as the decisions are made for us—quietly, invisibly, behind the tinted glass of a machine that does not need our hands, or our voice?

In revealing the illusion of arrival, we begin the necessary work of reckoning. Not just with the vehicles themselves, but with the structures—social, economic, political—that enable them. We learn to see not only the car on the road, but the networks that carry it, the values that guide it, and the futures it silently brings in tow.

From Prototype to Pavement

The moment a machine begins to navigate a public road without direct human control, it quietly declares that the boundary between experimentation and everyday life has been crossed. This line is easy to miss—not marked by ribbon cuttings or landmark legislation—but its consequences are far-reaching. The evolution of autonomous vehicles (AVs) from laboratory curiosities to operational presences in cities is not a sudden triumph of engineering. It is a cumulative process marked by strategic deployment, trial-and-error iteration, and a shifting interplay between public perception and corporate ambition. The self-driving car is no longer a speculative vision; it is already, in certain quarters, an embedded part of urban life. What remains elusive is a clear understanding of how we got here—and where this trajectory is leading us.

The earliest autonomous vehicle experiments were not sleek consumer-facing products but heavily outfitted research platforms—bulky, awkward, and constrained to closed courses. In the mid-1980s, Carnegie Mellon University's Navlab project and Mercedes-Benz's work with the Bundeswehr University in Munich laid early foundations, demonstrating that vehicles could follow basic lane markings or avoid obstacles. These prototypes were technical marvels for their time, but wholly impractical for public deployment. Their significance lay in showing that perception, decision-making, and control—functions previously assumed to require human cognition—could be translated into software. It was not autonomy in any practical sense, but rather an exploration of machine competence in tightly controlled conditions.

The real shift came when industry players entered the field with serious capital and a consumer-facing vision. In 2004, the U.S. Defence Advanced Research Projects Agency (DARPA) launched its first Grand Challenge, inviting teams to develop autonomous vehicles capable of navigating a desert course. The results were underwhelming—no team completed the route—but the event catalysed a surge of innovation. By the time the second and third DARPA challenges concluded, vehicles were completing complex routes with growing confidence. More importantly, these competitions attracted the attention of Silicon Valley, and a new narrative began to form: autonomy was not just possible, it was imminent.

That narrative has proven both motivating and misleading. Companies such as Google (later Waymo), Tesla, Uber, Cruise, and Baidu began to invest billions into AV research, framing it not as a long-term goal, but as a near-future disruption. The language used was often breathless. Timelines compressed from decades to years. Engineers were heralded as pioneers. Investors poured in capital. The race to release a fully autonomous car became a proxy for technological leadership. In doing so, the sector rapidly outpaced its own practical capabilities, creating a disjunction between public expectation and operational reality.

What distinguishes the present moment is not that AVs are flawless or widely adopted—they are not—but that they have become a lived presence. In select test zones, they pick up passengers, respond to traffic signals, and navigate urban complexity with increasing, if imperfect, precision. In cities like San Francisco and Phoenix, residents have reported seeing AVs make legal left turns, yield to pedestrians, and complete deliveries. These are not simulations. They are public performances, executed in real-world conditions and subject to the scrutiny of regulators, media, and local communities.

Yet the transition from prototype to pavement has not been uniform. Most deployments rely heavily on geofencing—restricted areas where the AV operates under predictable conditions. Weather, lighting, and road complexity are often managed by strategic route selection. In practice, this means the vehicle is not truly autonomous in the open-world sense. It is autonomous *enough* within a curated bubble. This partial rollout reflects a central tension: the desire to demonstrate progress while managing the risk of failure.

Such constraints are rarely visible to the casual observer. A vehicle cruising through downtown might appear to be self-reliant, but behind the scenes, it may be backed by remote operators, edge-case intervention protocols, and safety drivers ready to take over. The illusion of independence is carefully maintained because public trust depends on it. No company wants to advertise that its car can only function under optimal conditions. But those are the conditions under which many AVs currently operate.

Urban environments add further complexity. Unlike controlled test tracks, cities are sites of constant improvisation—construction zones, cyclists veering into traffic, pedestrians glued to their phones, delivery vans double-parked without warning. These situations demand not just perception but interpretation. Human drivers excel at reading intent: the jogger waiting to cross, the taxi driver about to cut across lanes, the elderly pedestrian hesitant at the kerb. Machines process visual inputs with growing sophistication, but nuance remains elusive. Context is hard to code.

Still, AVs are altering urban realities, even in their limited form. In test cities, municipal infrastructure is beginning to adapt. Road markings are made clearer to accommodate AV sensors. Kerbs are redesigned to facilitate safe pick-up zones. Dynamic signage and wireless communications are being integrated to synchronise with onboard systems. This is not passive accommodation—it is active reconfiguration, and it reshapes the spatial politics of the city. The machine's needs become the template for design.

This has subtle consequences. Streets evolve not around human messiness, but around machine expectations. Pedestrian unpredictability becomes a problem to be engineered out. Eye contact, a key part of non-verbal negotiation in traffic, becomes irrelevant. Urban planners, once primarily concerned with human flow and aesthetics, now face pressure to optimise for sensor visibility and computational predictability. The AV doesn't merely adapt to the city; it prompts the city to adapt to it.

One of the most significant shifts lies in the operational logic of the vehicle itself. Traditional cars are reactive: they respond to inputs from the driver. AVs are anticipatory, built to forecast and plan based on probabilistic modelling. This inversion has implications for liability, regulation, and the very meaning of driving. When a crash occurs, fault is no longer assigned to a person in a moment of distraction—it must be traced through logs, decision trees, sensor calibration, and machine learning inference. The site of responsibility moves from the roadside to the server room.

Notably, public exposure to AVs does not guarantee public understanding. Few passengers in a self-driving taxi are aware of how it perceives space, how it makes decisions, or what its limitations are. Companies have little incentive to disclose the inner workings of their systems. Intellectual property protection and reputational risk discourage transparency. As a result, there is a growing asymmetry between those who build the systems and those who use—or are impacted by—them. The ride may be smooth, but the governance is opaque.

Contrasting national strategies highlight this divide. In the United States, AV development has largely been led by private firms with piecemeal regulatory oversight. This approach fosters innovation, but often leaves cities scrambling to keep pace. In contrast, China's government-led model integrates AVs into broader urban development strategies, using them as instruments of both mobility and surveillance. In Germany, the emphasis has been on safety certification and ethical review, resulting in slower but more cautious deployment. These differences matter because they shape not just when and where AVs operate, but under what values and expectations.

Beneath these strategic contrasts lies a shared reality: the path from prototype to pavement is not linear. It is recursive, shaped by public trials, technical hiccups, media narratives, and legal entanglements. Each collision, each software update, each municipal hearing leaves a trace that influences the next step. The development of AVs is less like a product launch and more like a constitutional amendment—slow, contested, and contingent on public consensus.

There are, of course, commercial imperatives at play. The AV is not being built solely for its technical challenge or societal benefit. It is being built because it promises to reconfigure value chains in transport, logistics, and data. Companies envision fleets of AVs not as individual products sold to customers, but as networked assets generating continuous revenue. This model favours scalability, surveillance, and standardisation. The AV is not just a car without a driver—it is a node in a monetisable system.

Understanding the implications of this transition requires more than enthusiasm for innovation. It requires a sober assessment of how AVs are being introduced, who is making the decisions, and what assumptions are embedded in the technology. From a distance, the path from prototype to pavement may look like a victory lap. Up close, it reveals a mosaic of cautious steps, political negotiation, and design trade-offs. The self-driving car has arrived, but only in fragments—and those fragments are already reshaping the world around them.

In recognising this, we are better equipped to ask the questions that matter: Are AVs solving the right problems? Are they being deployed in ways that reflect public interest or private convenience? Are the cities they are helping to create ones we want to live in? The answers to these questions are not embedded in algorithms. They are political, cultural, and deeply human. And they begin with clarity—about where we truly are, and how we got here.

Technological Mythologies

In 2015, a senior executive at Tesla confidently predicted that within two years, a car would be able to drive autonomously from Los Angeles to New York without human intervention. Nearly a decade later, that coast-to-coast journey remains unrealised—not for lack of effort, but because the problem is vastly more complex than the early promises allowed. And yet, public faith in the inevitability of self-driving cars remains remarkably resilient. This resilience is not sustained by technical proof, but by narrative force. Silicon Valley has not merely built machines—it has built mythologies. These mythologies frame technology as destiny, collapse timelines into digestible talking points, and reduce moral ambiguity to engineering challenges. They render the messy, uncertain reality of autonomous vehicles into a story of inevitable progress, where the future is always just around the next software update.

This myth-making is not a side effect of innovation; it is a strategic asset. It secures investment, attracts talent, and shapes regulatory goodwill. The image of the driverless car as a frictionless, safe, and democratising force is rehearsed with almost liturgical precision at every product launch, TED talk, and earnings call. The narrative is familiar: human drivers are fallible; machines are precise. Roads are dangerous; automation will make them safe. Cities are inefficient; AVs will make them smart. In this schema, resistance is not debate but delay. The story's momentum discourages scrutiny and punishes hesitation. Yet when examined closely, the edifice of inevitability begins to show strain—not because the ambition is misguided, but because the narrative conceals as much as it reveals.

To understand the endurance of these technological mythologies, one must consider the culture that produces them. Silicon Valley has long operated on a set of assumptions that blur the line between prediction and prescription. Chief among them is the belief that *if something can be built, it should be.* The moral calculus is built not on social deliberation but on engineering potential. In the context of AVs, this translates into a peculiar inversion: instead of asking whether the public wants or needs driverless cars, companies build them and then seek to retrofit public desire. The vehicle is not a response to demand; it is an attempt to create it.

This inversion is particularly stark when viewed against the infrastructural and ethical realities AVs confront. Machine learning systems that govern AV behaviour are trained on vast datasets—but these datasets reflect existing societal patterns, including racial bias, spatial inequality, and infrastructural gaps. An AV that is "trained" in Palo Alto may falter in Detroit or Dhaka—not just because of road conditions, but because the underlying logic assumes a narrow band of cultural and infrastructural norms. Yet in the mythic framing, these differences are not substantive, only temporary. The world is implicitly expected to adapt to the AV, not the reverse.

One of the most persistent fictions is that self-driving technology is a clean break from the past. In truth, it is deeply entangled with historical legacies—particularly the American obsession with car-centric urban design. The AV is often sold as a disruption, but it fits neatly into the existing paradigm of automotive dominance. It does not challenge the primacy of cars over public transport, cyclists, or pedestrians; it extends and optimises it. The fantasy of point-to-point, on-demand mobility without human labour is not a rethinking of mobility—it is a refinement of a model that has already produced environmental degradation, spatial inequality, and social isolation. The mythology hides this continuity beneath the promise of novelty.

The ethical dilemmas embedded in AV systems are similarly abstracted or deferred. When questioned about moral decision-making—such as the infamous "trolley problem"—AV developers often respond with pragmatism: real-world decisions, they argue, are made probabilistically, not philosophically. But this deflection avoids the core issue: *whose values are encoded into the machine, and who gets to decide?* Silicon Valley's mythology treats value conflicts as bugs to be fixed rather than as questions requiring democratic engagement. This evasion is not incidental—it is central to the myth's function. Complexity must be simplified, contingency must be minimised, and dissent must be reframed as irrational fear of progress.

Public trust, meanwhile, is maintained through selective transparency. Companies publish glossy safety reports, but omit critical incident data. They court media coverage with staged "driverless" rides that still involve remote supervision. The mythology relies on controlled exposure: enough visibility to maintain intrigue, but not enough to provoke disillusionment. This curation is often internalised by users themselves, who extrapolate from brief, seemingly flawless experiences to assume maturity and safety. The occasional crash is framed as an anomaly rather than a structural concern. Here, the mythology overlaps with the culture of convenience—if a ride feels seamless, it is easy to believe the system must be sound.

Consider the way in which technological milestones are announced. Companies herald the "first fully driverless ride" or "safety driver-free pilot," even as these operations occur under tightly constrained conditions. Geofencing, weather limitations, and pre-mapped routes are downplayed. The complexity of real-world driving is rendered as a checklist of obstacles already overcome. In doing so, the industry creates a perception of arrival that precedes its reality. This perception has real consequences. It shapes investor behaviour, guides policymaking, and dulls the edge of public inquiry.

Contrasting examples from other sectors reveal just how particular this myth-making is. Commercial aviation, for instance, operates under rigorous safety regimes, public transparency requirements, and deeply institutionalised oversight. When a plane crashes, investigators reconstruct every detail, and systemic improvements follow. In AV development, by contrast, crashes often lead to vague statements and opaque investigations. The mythology shields the industry from the accountability standards expected of other transport systems. The rationale is that AVs are still learning—and that, in time, the machines will exceed human performance. But that promise, endlessly deferred, begins to resemble faith more than evidence.

Faith, of course, is not inherently misplaced. The aspiration to build safer, more accessible, and less labour-intensive transport is noble. But aspiration without scrutiny produces systems that are optimised for the wrong goals. In the AV sector, optimisation often means reducing friction—legal, physical, social—in the name of efficiency. But friction is not always a flaw. It is sometimes the texture of democracy: the slow, deliberate negotiation of competing interests. When technological mythologies paper over these frictions, they do not eliminate them—they displace them. Risk is shifted to edge cases, to marginal users, to cities unequipped to resist corporate terms of deployment.

This asymmetry is compounded by the global export of Silicon Valley's assumptions. The AV narrative, born in a narrow cultural and geographic context, is now repackaged for international markets with little adjustment. Countries with different regulatory cultures, public transport traditions, or spatial patterns are offered the same vision: a world where machines take the wheel and humans are freed from the burden of driving. But this is a deeply American ideal, rooted in the history of the open road and the illusion of limitless mobility. Transplanted elsewhere, it can clash with urban density, collective transport models, or data sovereignty norms. The mythology, however, resists localisation. It travels easily because it is abstracted from place, context, and constraint.

A more honest discourse about AVs would acknowledge uncertainty as a feature, not a flaw. It would resist the urge to narrate progress as a straight line. It would distinguish between ambition and reality, between what machines *can* do and what they *should* be asked to do. It would invite the public not just to witness the rise of AVs, but to shape their role. And it would treat ethics, infrastructure, and governance not as technicalities to be solved post-deployment, but as foundational design questions.

Technological mythologies thrive on silence—the silence of systems running smoothly, of unanswered questions, of unexamined trade-offs. Breaking that silence is not a matter of cynicism. It is a matter of responsibility. To participate in the future of transport requires more than passive observation. It requires the courage to ask: *What are we being promised? What are we not being told? And who benefits from the story we're being asked to believe?* The mythology of the autonomous vehicle is powerful precisely because it is incomplete. Completing the picture requires more than engineering. It requires an act of collective vision—and the willingness to speak plainly in the face of seductive futures.

Why 'Self-Driving' Is a Misnomer

No car drives itself. However sophisticated the system, however smooth the ride, however absent the human hands on the wheel, the notion that a vehicle is operating truly independently is an illusion—one sustained by marketing ambition, reinforced by media shorthand, and rarely interrogated in public discourse. The term *self-driving* implies an autonomy that does not exist. It suggests that a machine can replace not only human input but also the broader web of dependencies—legal, technical, infrastructural, ethical—that make driving possible in the first place. This terminology is more than misleading; it obscures the nature of the systems we are adopting and misplaces the burden of responsibility in ways that could carry lasting social and political consequences.

The seductiveness of the term lies in its apparent simplicity. Like "wireless" or "contactless," it conceals the extraordinary complexity required to produce the outcome it describes. But where those terms refer to the removal of physical connections, *self-driving* gestures toward the removal of human agency—and that is a far more loaded proposition. What is being claimed, explicitly or otherwise, is that machines can independently perceive, decide, and act in environments historically navigated by humans, under conditions of risk and uncertainty. The reality is that every so-called self-driving car is deeply reliant on an interlocking set of human-made and human-managed frameworks: data pipelines, software engineers, infrastructure planners, remote operators, policy negotiators, and regulatory bodies. The autonomy, such as it exists, is partial, provisional, and heavily conditional.

Consider the basic requirement of perception. An AV must detect and interpret its surroundings with a level of accuracy that approximates—or, ideally, exceeds—human capability. But it does not do this by seeing the world as we do. Instead, it interprets sensor inputs—LIDAR, radar, ultrasonic data, camera feeds—through probabilistic models built on training datasets. These datasets are curated and labelled by human workers, often underpaid and anonymised in outsourced data-labelling operations across the Global South. The vehicle does not learn by experience in the human sense. It learns by pattern recognition derived from countless hours of human-supervised data ingestion. When it recognises a stop sign, it is not seeing it as a symbol within a cultural system of rules; it is matching visual input to statistical likelihoods provided by engineers who have defined what "a stop sign" is. The illusion of self-sufficiency begins to crumble under the weight of these upstream dependencies.

Navigation presents another layer of reliance. No AV is operating in a vacuum. Each depends on high-definition maps, often updated in real time, to provide centimetre-level resolution of the driving environment. These maps are created not by the vehicle itself but by a complex process of spatial data acquisition, aggregation, and verification. Any change in road layout, signage, or construction must be swiftly integrated into this system to avoid failure. Mapping, far from being a solved problem, is an ongoing, labour-intensive endeavour that links the AV to a global network of infrastructure and human maintenance. The car does not "know" where it is; it is constantly being told.

Nor can an AV operate without the institutional scaffolding of regulation and governance. Driving is not merely a mechanical act—it is a legally codified behaviour that intersects with policing, liability, insurance, and public accountability. A machine that moves on public roads must comply with traffic laws, which themselves vary by jurisdiction and evolve over time. When an AV encounters an ambiguous or novel situation—say, an unexpected police roadblock or an elderly person crossing the street outside a marked zone—it must make a decision that is both technically feasible and legally defensible. These are not binary outcomes. They involve judgement, priority-setting, and trade-offs. No machine can navigate these dilemmas without being preloaded with normative assumptions—assumptions made by human programmers, often behind closed doors.

And yet, the term *self-driving* implies that the car is, in effect, a moral actor. It grants the vehicle a false subjecthood, a misleading identity that subtly shifts responsibility away from its creators. When a self-driving car makes a poor decision, media reports often describe it as the car's mistake, not the system's. This narrative displaces accountability and reinforces the myth that machines are making their own choices, when in fact those choices are the product of human design. It also creates a dangerous precedent: as machines are increasingly personified, they are also increasingly insulated from the kinds of scrutiny we demand of human drivers. The designation of "self-driving" turns a collaborative, socio-technical system into a black box of blame avoidance.

The political implications of this misnomer are significant. Public trust in AVs is being cultivated on the premise of independence—on the idea that, given enough testing, enough data, enough refinement, these machines will become fully capable, standalone agents of mobility. But if the foundational claim is flawed, the trust being built upon it is similarly brittle. The vehicle may appear autonomous to its user, but this autonomy is contingent on constant human involvement behind the scenes. That includes the engineers who write the code, the emergency remote operators who intervene when a vehicle becomes stuck or confused, the urban planners adjusting kerb heights and lane widths to make AV navigation easier, and the legal teams shaping the parameters of acceptable liability. Each "self-driving" moment is underwritten by a distributed, invisible labour force.

This discrepancy between surface perception and systemic reality is not merely academic. It affects policy, public discourse, and legal infrastructure. Legislators tasked with regulating AVs often lack the technical fluency to distinguish marketing language from operational constraints. They may assume that autonomy implies independence, and thus adopt a hands-off approach to oversight—believing, incorrectly, that once the car is on the road, the human element has been designed out. In truth, human involvement has simply shifted from the wheel to the cloud, from the driver's seat to the command centre. Regulation that ignores this redistribution is regulation that fails to account for where responsibility actually resides.

Moreover, the fantasy of full autonomy diverts attention from more urgent and solvable problems. Public investment and policy energy are often funnelled into supporting AV rollouts at the expense of more equitable and immediately beneficial transportation solutions—such as expanding bus networks, investing in pedestrian infrastructure, or subsidising accessible transport for the elderly. The term *self-driving* seduces with the promise of a frictionless future, but it also distorts collective priorities. It suggests that the car is the problem to be perfected, rather than the system to be re-evaluated.

There is also a cultural cost. The language of autonomy subtly redefines what it means to drive, to move, to navigate space. It transforms mobility from an act of participation into a service to be consumed. It removes the idea of co-presence and mutual recognition among road users and replaces it with a transactional interaction between user and platform. The "self" in *self-driving* is not the human being—it is the machine. And that machine is designed not only to transport but to predict, to record, and to optimise. Its sensors see more than the road. They see you: your habits, your routes, your stops, your hesitations. To ride in a so-called self-driving car is to enter a system of continuous surveillance, masked by a promise of liberation.

A more honest vocabulary would reflect the reality of dependence. Terms like "automated driving system," "machine-assisted vehicle," or "networked autonomy" might lack the rhetorical punch of *self-driving*, but they would anchor the conversation in what these systems actually are: *interdependent*, *human-shaped*, and *context-sensitive*. The move toward semantic precision is not merely pedantic. It is political. Language defines expectations. Expectations shape regulation. And regulation determines the conditions under which these machines will—or will not—serve the public good.

To interrogate the term *self-driving* is not to reject the project of automation. It is to insist that we understand what we are building, what we are calling it, and who is implicated in its operation. The road ahead is not driverless. It is populated—by systems, by labour, by choices, and by consequences. The car may no longer need your hands, but it is still guided by human intent. The autonomy we speak of is, at best, partial. At worst, it is a rhetorical sleight of hand. And unless we call it what it is, we risk building a transport future on the scaffolding of a linguistic myth.

The Culture of Convenience

A machine that arrives at your doorstep on demand, silently whisks you across town while you read, eat, or sleep, and deposits you without a word at your destination—this is the version of the future that autonomous vehicle manufacturers are selling, and it resonates because it speaks to something deeper than technological enthusiasm: a cultural entrenchment in convenience. In consumer economies, convenience is often treated not as a preference, but as a right—a quality so deeply woven into modern life that its absence feels like a flaw in the system. It is no surprise, then, that convenience has become the dominant design principle for everything from digital platforms to physical infrastructure. What is less examined is how this appetite for ease distorts our ability to assess trade-offs—

particularly when it comes to privacy, safety, and civic accountability. The driverless car is not merely a technological artefact; it is a crystallisation of our collective willingness to overlook structural risks in exchange for momentary relief from friction.

Convenience is rarely questioned because it presents itself as self-evidently good. It saves time. It simplifies decision-making. It removes burdens. Yet these very qualities make it difficult to interrogate. When we are offered more convenience—less waiting, less effort, fewer steps—the proposition seems almost morally neutral. But convenience is not distributed evenly, nor is it politically inert. It has a direction, a beneficiary, and often, an invisible cost. Autonomous vehicles amplify these dynamics by embedding them into systems that function largely without user visibility. The convenience of a car that picks you up automatically, maps the fastest route, and requires no conversation is not just a user experience—it is a framework for reshaping mobility, data collection, and the built environment in ways that privilege corporate logic over public deliberation.

Take, for instance, the role of data. To make driving feel effortless, AVs must continuously gather and process vast quantities of information: environmental inputs, pedestrian patterns, road conditions, passenger identities, biometric markers, and behavioural trends. Each layer of convenience—personalised routes, predictive services, seamless payment—requires a deeper layer of surveillance. This data is not discarded once the ride is complete. It is stored, analysed, and often commodified, becoming part of a behavioural profile that can be used for pricing, marketing, or even law enforcement. The trade-off here is clear but frequently unstated: to be relieved of friction, you must become transparent to the machine.

What's remarkable is how rarely this surveillance is perceived as intrusive. Much of this has to do with framing. When data collection is bundled with convenience, it becomes difficult to reject. Few users pause to question how their ride was so seamless, or who benefits from the data generated. The experience is engineered to produce a feeling of competence and trust: the machine knows what you need, where you're going, and how to get you there faster than you could on your own. The system works so well that it becomes easy to forget you are inside a transaction—not only of services, but of information, power, and responsibility.

Public safety is similarly reframed under the logic of convenience. AV companies regularly cite human error as the leading cause of road fatalities—an often-repeated statistic used to justify automation as a safer alternative. While the claim holds statistical weight, it also obscures a crucial tension. Safety, in the context of AVs, is defined largely in technological terms: reduction in collisions, improvement in reaction time, consistency of operation. But safety is also a social condition, shaped by context, trust, and access. A vehicle that avoids accidents but marginalises pedestrians by rerouting them for machine efficiency is not enhancing safety in any holistic sense—it is redefining it to suit system parameters. Similarly, a transport system that monitors every movement to prevent crime may indeed deter certain behaviours, but it does so at the expense of civil liberties. The convenient route is not always the just one.

The language of ease also tends to flatten distinctions that are ethically significant. For instance, the difference between user control and user dependency becomes difficult to detect when both are dressed in the same interface. A rider who uses an AV service daily may believe they are exercising choice, but over time, patterns of availability, pricing, and algorithmic nudging can subtly reduce that freedom. When a route is chosen for you, a service area geofenced, or a destination deprioritised, the experience still feels seamless—but it has quietly narrowed the boundaries of autonomy. The comfort of not having to choose can, paradoxically, become a form of engineered passivity.

Historical parallels make these dynamics easier to grasp. The rise of the automobile in the 20th century promised unparalleled freedom and flexibility. What followed was suburban sprawl, environmental degradation, and the dismantling of public transport networks. Each of these consequences was enabled, in part, by a public willing to exchange complexity for convenience. The self-driving car, despite its futuristic veneer, follows the same arc: it promises more freedom while deepening dependence—on platforms, on data infrastructures, and on opaque corporate governance. The danger is not that we will lose control overnight, but that we will give it away piece by piece, each time the trade-off seems too minor to resist.

This incremental surrender is rarely visible because it is embedded in design. AV interfaces are built to be intuitive, seamless, even pleasurable. Every friction point removed is a moment of micro-relief—no need to hail a taxi, remember directions, or argue over pricing. These small wins accumulate into a user experience that feels empowering. But beneath that experience lies a set of architectural decisions about who is visible to the system, who is prioritised by the algorithm, and who is left out entirely. For example, areas deemed high-risk or low-profit may be excluded from AV service zones, reinforcing spatial inequalities under the guise of optimisation. Riders in these areas are not choosing inconvenience—they are being denied convenience by design.

There is also an intergenerational aspect to consider. Young users raised in platform ecosystems—where transport, food, communication, and entertainment are all one tap away—are habituated to systems that anticipate rather than ask. For them, the expectation is not just ease, but invisibility of effort. This expectation is not value-neutral. It shapes how civic engagement, public goods, and collective responsibilities are perceived. If convenience becomes the baseline for legitimacy, anything that requires negotiation, deliberation, or compromise is seen as inefficient. Public transport delays, pedestrian detours, and infrastructure investments are framed not as necessary components of shared life, but as annoyances to be engineered out.

To interrogate the culture of convenience, then, is not to reject efficiency or technological advancement. It is to ask what kind of society we are optimising for. Who decides what is worth simplifying, and at what cost? What values are embedded in systems designed to serve us with minimal input? And how do we ensure that ease of use does not become ease of manipulation?

One approach is to demand transparency—not just in how AVs function, but in the assumptions driving their design. Users should know how their data is used, how decisions are made, and what rights they retain—or relinquish—when they step into a driverless car. Regulators should treat convenience as a domain of governance, not merely a feature of innovation. And designers should recognise that friction is not always a flaw. Sometimes, it is the very thing that reminds us we are still participants, not just passengers.

Convenience, in its highest form, should expand freedom, not replace it. It should make life more navigable without making us more manageable. The challenge with AVs is that they offer a seductive form of convenience—one that relieves us of effort while binding us more tightly to systems we do not control. Unless we examine that dynamic clearly, we risk building a future in which mobility is effortless, but agency is elusive. The ride may be smooth, but the cost of not asking where it's taking us may be far higher than we imagine.

Chapter 2
Who Holds the Wheel Now?

Driving has always been synonymous with control. To grip the wheel, choose a route, and navigate uncertainty is to exercise agency—to act, decide, and accept responsibility. For over a century, the driver has been central not only to the vehicle's function but also to the social and legal frameworks surrounding it. Yet as autonomy seeps into the space once occupied by human reflex and intuition, this notion of control is quietly dissolving. The wheel remains, but its meaning shifts—from a site of decision to a placeholder within a vast, abstract network of systems, algorithms, and administrative proxies. The pressing question is no longer how we drive, but who—or what—now governs movement when no hand is visibly at the helm.

This chapter begins by confronting that ambiguity. Control, once tactile and singular, is now dispersed across codebases, backend infrastructures, and multi-party contracts. Autonomy, presented in glossy promotional videos as seamless, frictionless operation, belies a more complex reality: the locus of authority is scattered, opaque, and often inaccessible to those inside the vehicle. Control does not vanish; it becomes harder to locate, harder to contest, harder to comprehend. This shift is not merely operational—it alters our relationship with the very systems that transport us.

Historical precedent offers some guidance. Computers replaced manual calculations, GPS systems supplanted mental navigation, automated checkouts assumed transactional tasks. Each promised efficiency while introducing new dependencies. What distinguishes AVs, however, is the scale and stakes: here, we are surrendering an entire dimension of behavioural autonomy. Driving encompasses ethics, negotiation, situational nuance, and interpersonal cues. Delegating the wheel means outsourcing these embedded judgements—often to code developed thousands of miles away, tested in unfamiliar conditions, and owned by actors whose priorities may not align with the public good.

Design plays a pivotal role in mediating this shift. Touchscreens, voice prompts, and app-based interfaces are far from neutral—they shape perception, behaviour, and expectations. A passenger may set a destination but not a route, pause the ride but not override an emergency stop. These constraints are framed as convenience or safety, masking the trade-off: diminished agency, restricted oversight, and reduced transparency. The interface functions as both gatekeeper and disguise.

This deliberate removal of human discretion reflects a broader cultural trend. Systems are increasingly built to replace decision-making rather than assist it, reframing risk, responsibility, and trust. When failures occur—and they inevitably will—assigning fault becomes labyrinthine. Traditional legal frameworks rely on individual culpability. In autonomous contexts, blame fragments: sensor error, mapping misclassification, software oversight, fleet management, or municipal approvals—all potential contributors. Each requires layers of technical and legal interpretation, distancing accountability from individuals toward institutions often shielded from scrutiny.

The 2018 tragedy involving Elaine Herzberg, struck by an Uber self-driving test vehicle in Arizona, exemplifies this challenge. The safety driver was charged with negligence, despite the software failing to identify Herzberg as a pedestrian. The incident revealed a profound misalignment between legal norms predicated on individual responsibility and the distributed reality of machine-driven decision-making. The wheel was present, the driver was present, yet conventional control seemed absent.

Ownership, too, is being redefined. AVs are frequently paired with mobility-as-a-service (MaaS) models, shifting users from owners to subscribers. While seemingly efficient, this transformation alters the individual's relationship to mobility. Ownership confers rights: to modify, maintain, or refuse use. Subscription models define rights through terms of service—revocable, updateable, asymmetrical. Access, now conditional on algorithms, pricing, location, or behavioural profiling, replaces possession. Mobility becomes a privilege contingent on compliance, rather than a fundamental autonomy.

Together, these shifts—of control, responsibility, and ownership—signal a profound recalibration of the transport contract. AVs do not merely remove the driver; they obscure who is accountable, who governs, and who benefits. For designers, engineers, and policymakers, this is as much a philosophical question as a technical one: how do we trust systems that cannot be interrogated? What recourse exists when harm stems from code buried in proprietary platforms? How do we prevent ease of use from becoming ease of exclusion?

Addressing these challenges requires moving beyond a simplistic pro- or anti-technology stance. The question is not whether AVs are inherently good or bad, but whether we understand—and are prepared to govern—the systems they instantiate. Control, once tangible, is now distributed through code, law, capital, and interface. Responsibility, once assignable, flows across networks of design and delegation. Ownership, once a source of agency, is being replaced by conditional access.

This chapter does not romanticise the human driver, who has long posed risks and inefficiencies. Rather, it recognises that replacing the human does not erase complexity—it reconfigures it. Unless we confront that reconfiguration, we risk inhabiting a system where the wheel still turns, yet no one can discern who holds it, to whom it answers, or where it is taking us. The road ahead is being drafted now—in code, contracts, policies, defaults, and assumptions. Our task is not merely to ride along, but to ask, with clarity and urgency: who holds the wheel today, and what would it take to reclaim meaningful agency in a driverless world?

Shifting Definitions of Control

A person behind the wheel once epitomised control in its most immediate, physical form: the ability to steer, brake, accelerate, and make decisions based on perception, judgement, and situational awareness. That control was not merely mechanical—it was symbolic. It implied ownership, accountability, and presence. The driver was the legal agent, the ethical actor, the node of responsibility. Yet the rise of autonomous vehicles is dissolving that clarity. Control, once singular and legible, is now diffused across layers of software, infrastructure, corporate governance, and remote oversight. What emerges is a system that functions with increasing independence from human input, while simultaneously relying on an intricate mesh of distributed decision-making. Understanding this evolution is essential—not only for interpreting how AVs operate, but for grasping the shifting contours of agency, accountability, and power in a machine-mediated world.

The human driver is being replaced not by another individual, but by a system—a distributed architecture of algorithms, sensors, cloud services, and policy frameworks. This system is designed to simulate, and in some respects surpass, the capabilities of human cognition. It processes vast streams of environmental data in real time, maps millions of potential trajectories, and adjusts behaviour based on predictive modelling. Yet this sophistication comes at the cost of transparency. Unlike a human driver, who can explain their reasoning ("I braked because the cyclist swerved"), the AV's "decision" is an emergent property of code trained on statistical probabilities, running across machine learning models whose inner workings are often inscrutable even to their creators. Control, in this context, no longer belongs to a person—it resides in the logic of the system.

This system, importantly, is not contained within the vehicle. It stretches outward. The so-called autonomous car is tethered to data centres, reliant on remote software updates, dependent on up-to-date high-definition maps, and often monitored by human supervisors who can intervene remotely if something goes wrong. These layers form what might be called a *stacked control system*—a layered hierarchy in which decision-making is distributed across time, space, and institutional actors. The vehicle reacts in real time, but only within constraints established far earlier: by the engineers who wrote the code, the corporations who set operating policies, the governments who defined compliance protocols, and the service models that determined use cases.

This decentralisation has profound implications. It severs the link between interface and authority. A user might sit in the driver's seat—or more likely, the passenger seat—and feel in control because they can select a destination or cancel a trip. But these inputs exist within a bounded system. The user cannot interrogate the route choice algorithm, override a machine's decision to reroute, or negotiate with the vehicle in the way they might with a human taxi driver. Control, in this sense, is not exercised—it is *simulated.* The interface offers the illusion of autonomy, while insulating the real decision architecture from user influence.

The legal system, historically built around the premise of identifiable agency, struggles to keep pace. Liability frameworks assume a driver—or at least an operator—who can be held responsible when something goes wrong. But in a distributed system, fault becomes a forensic problem, not a social one. Who is responsible when an AV fails to yield at a crosswalk: the manufacturer of the braking system, the software engineer who tuned the object detection model, the mapping team who failed to update local road geometry, or the remote supervisor who hesitated to intervene? Each answer leads to a different layer of the stack, and none points to a single, accountable figure. The practical result is a dilution of responsibility. It becomes harder to assign blame, harder to extract justice, and harder to maintain public trust in the face of failure.

Compounding this is the question of recourse. In a conventional system, a driver can contest a ticket, appeal a fine, or argue mitigating circumstances. But when control is algorithmic, decisions are often final, their rationale hidden behind proprietary code and closed-loop logic. If an AV denies a route, slows unexpectedly, or performs an emergency stop, the human occupant may have no meaningful way to contest or even understand that action. Control, once exercised through feedback and response, becomes asymmetrical. The system acts; the human complies. The absence of explanation reinforces the asymmetry, making it difficult to assert rights or challenge outcomes.

The asymmetry is not only between human and machine, but also between users and system designers. Those who write the code, define the parameters, and control the update mechanisms have vastly more influence over vehicle behaviour than those who ride in them. Yet they operate at a distance—geographically, institutionally, and often legally. The AV's logic is shaped in boardrooms and labs, not in the communities where it is deployed. This separation introduces a democratic deficit. Local contexts—cultural norms, infrastructural quirks, regional regulations—are often subsumed under a standardised logic of operation. The same system that performs flawlessly in one city may behave erratically in another, not because of technical fault, but because the model of control does not account for social nuance.

Consider the case of jaywalking. In many Western cities, pedestrians crossing against the light is common and socially accepted, even if technically illegal. Human drivers often accommodate this behaviour, adjusting speed or gesturing to indicate right of way. An AV, trained to obey traffic laws rigidly, may fail to recognise this as part of local driving etiquette. It may stop abruptly, reroute unnecessarily, or confuse hesitant motion for erratic behaviour. The system performs correctly by design, but incorrectly in context. Here, control has not failed technically—it has failed *relationally*. The vehicle is operating in isolation from the norms of its environment, and the person most affected—the pedestrian—has no access to the logic dictating the machine's behaviour.

The same pattern appears in vehicle-to-infrastructure (V2I) systems. As AVs become integrated with smart traffic lights, road sensors, and municipal data networks, their behaviour is increasingly shaped by infrastructure-level directives. The vehicle may prioritise routes based on city congestion algorithms, environmental impact calculations, or emergency response protocols. These choices may benefit the system as a whole, but they further distance control from the individual occupant. What appears to be a personalised ride is actually a computation across layers of competing priorities—many of which remain invisible to the user. Control becomes an emergent property of a networked negotiation, not a personal act of will.

At the extreme, this model raises a philosophical question: if control is everywhere and nowhere, what does it mean to be a participant in a system of movement? Are we users, stakeholders, data points, liabilities, or passengers in someone else's machine? The answer, increasingly, is all of the above. The AV ecosystem turns mobility into a *mediated experience*, one in which agency is distributed, contingent, and often conditional. This is not inherently negative. Distributed systems can be more efficient, more adaptive, and even more inclusive. But without visibility into how decisions are made and who gets to shape them, distributed control easily becomes unaccountable control.

What is needed, then, is a rethinking of control that aligns with the socio-technical reality of AV systems. This rethinking should recognise that control is no longer a static possession—it is a negotiated function, exercised through access, transparency, and recourse. It must account for the multiple actors who shape vehicle behaviour, the institutions that govern deployment, and the interfaces that frame user interaction. And it must embed mechanisms for contestation, adaptation, and public oversight. Without such a redefinition, we risk living in a transport future where the machine moves perfectly, but no one knows whose logic it follows, whose interests it serves, or how to speak back when it errs.

The shift in control from individual drivers to distributed systems is not simply a technical development. It is a constitutional transformation in the governance of mobility. The challenge is not to restore a past model of control, but to construct a new one—fit for the complexity of autonomous systems, and accountable to the societies in which they operate. If we fail to do so, we may find ourselves increasingly moved by forces we cannot see, in vehicles we cannot challenge, on roads that no longer belong to us.

Designing Out the Human

In the modern vehicle, the steering wheel is no longer just a tool of control—it is increasingly a relic of interface, a symbolic remnant of a role that is being quietly engineered out. As automation advances, human agency is not eliminated with a single stroke; it is *eroded by design.* Each new iteration of autonomous functionality—automatic braking, lane-keeping assistance, adaptive cruise control—removes one more layer of human involvement under the guise of safety, efficiency, and user experience. But the deeper transformation is less about what the human no longer does and more about what the machine now decides. The design of autonomous systems, especially in transport, is not simply technical—it is ideological. It encodes assumptions about what matters, what can be predicted, and who—or what—should be in charge. When designers claim to simplify the user experience, they are often reconfiguring the space in which human decision-making is possible at all.

To understand how this process unfolds, consider the evolution of the driver's role. In legacy vehicles, control was mechanical and immediate: brakes responded to foot pressure, wheels to hand motion, direction to visual cues. Errors were part of the human condition, and corrections were expected. But in automated vehicles, control is gradually absorbed by layers of logic designed to pre-empt, override, or isolate human behaviour. A lane-departure warning system, for instance, may vibrate the wheel if you drift. A more advanced system may gently guide the car back into lane. At a certain threshold, the car no longer alerts you—it simply acts. And in doing so, it makes a value judgement about what should happen, regardless of your intent. *The system does not negotiate with you; it interprets you*—as deviation, as signal noise, as risk to be managed.

This shift is not inherently problematic. In many cases, automated systems outperform human drivers in reaction time, attention span, and rule adherence. But the framing of automation as inherently superior misses a crucial distinction: *competence is not the same as contextual intelligence.* Humans may make mistakes, but they also improvise, signal, empathise, and adapt in ways that systems cannot. A driver might pause for a child on a scooter before it becomes a threat, or gesture to a pedestrian to cross despite having the legal right of way. These are not just reflexes—they are social behaviours embedded in a matrix of mutual recognition and situational awareness. When automation designs the human out of these moments, it strips the interaction of that social texture. The decision is not wrong—it is *contextually blind.*

The interface through which users engage with autonomous systems reflects this underlying logic. As automation deepens, interfaces are not expanded to accommodate shared decision-making; they are streamlined to support machine autonomy. In the name of frictionless experience, options are narrowed, controls simplified, feedback reduced. The assumption is that fewer choices mean better usability. But fewer choices also mean fewer opportunities to dissent, to modify, or to challenge. Consider a user who wishes to take a scenic route instead of the algorithmically optimal one. In a human-driven taxi, this request can be negotiated. In a driverless vehicle governed by route optimisation software, it may not be possible at all—or may require a complex workaround. The user becomes not a driver, but a managed variable within a system designed to perform without them.

There is a subtle but important distinction between *user-centred design* and *user-subordinated design*. The former begins with human needs and builds technology in support of them. The latter begins with technical constraints and slots human needs into those parameters. Many AV systems today, particularly those developed within Silicon Valley's engineering-first paradigm, lean toward the latter. The user experience is prioritised, but within a rigid framework that privileges predictability, control, and standardisation. Human behaviour that falls outside these bounds—uncertainty, spontaneity, non-verbal negotiation—is treated as an obstacle to be engineered out. *In this way, the system becomes efficient by becoming intolerant of the human.*

One of the most visible consequences of this shift is *diminished override capacity*. In traditional vehicles, the driver is the ultimate authority; any system failure can be countered by direct intervention. In highly automated systems, this is often no longer the case. When an AV makes a decision—whether to reroute, to stop, or to accelerate—the occupant may have no means of reversal. Even in vehicles designed with manual fallback modes, the speed and complexity of the decision loop can make meaningful intervention unrealistic. A user may be *legally* empowered to act, but *functionally* excluded by system latency, control design, or information opacity. Control is not removed—it is rendered inaccessible at the moment it is most needed.

Real-world incidents highlight how dangerous this can become. In several fatal accidents involving Tesla's Autopilot system, investigations revealed that drivers had placed excessive trust in the system's capabilities. They were not negligent in the conventional sense—they were behaving in accordance with how the interface framed their role. The system encouraged them to "keep hands on the wheel," but also handled most driving functions with little apparent need for intervention. The result was a form of *passive displacement*: the human was still present but increasingly disengaged, lulled by design into a state of semi-responsibility. When a decision point arrived, the person was cognitively and procedurally unprepared to act.

Designing out the human also has cultural and psychological dimensions. Over time, systems that discourage engagement recondition users. Familiarity breeds not awareness, but habituation. The more seamless the ride, the less we interrogate its terms. This is particularly acute in younger users, who may come of age in systems where choice is structured not by preference but by platform defaults. Just as GPS has altered our sense of spatial orientation, AVs may alter our sense of responsibility and situational awareness. A user who never makes a decision begins to internalise the idea that decision-making is either unnecessary or inappropriate. *This is not laziness—it is learned dependence.* And it reflects a broader societal shift toward passive consumption of services that were once participatory.

Critics might argue that automation, by removing the human, also removes human error. But this is a false dichotomy. The issue is not whether humans make mistakes—they do—but whether removing them also removes something *essential*: context, ethics, adaptability, judgement. A well-designed AV system should not aspire to eliminate the human. It should *honour the human's strengths while compensating for their limitations.* Instead, too many systems today treat human deviation as the problem to be solved rather than a source of insight to be integrated. In doing so, they miss the opportunity for genuine augmentation—an approach in which machine and person operate not in succession, but in dialogue.

The challenge, then, is not to resist automation per se, but to insist on *meaningful human agency within automated systems.* This means designing for override, for explanation, for input. It means creating systems that can negotiate with users rather than instruct them. It means embedding flexibility rather than enforcing compliance. And above all, it means recognising that the human is not a weakness in the system, but a necessary counterweight to the blind spots of logic-driven design.

Designing out the human is not simply a design decision—it is a cultural trajectory. It reflects choices about values, about power, and about the future of decision-making itself. The danger is not that machines will overpower us, but that we will *cede authority without even noticing*—not because we trust the system, but because the system no longer asks us to be present. In that absence, autonomy becomes automation. And what is lost is not control in the mechanical sense, but *the capacity to act meaningfully within a system that was supposed to serve us*. If we wish to reclaim that capacity, we must design systems that not only perform for us, but partner with us—systems that expect us to think, to choose, and to matter.

Responsibility in the Rearview

When a human driver runs a red light and causes a collision, legal systems across the globe know what to do. There are established frameworks for investigating the event, assigning fault, penalising negligence, and compensating victims. The presence of a fallible but accountable driver provides a centre of gravity around which liability and responsibility can be organised. That gravitational pull disappears in an autonomous vehicle crash. The scene of the accident may look familiar—bent metal, flashing lights, forensic tape—but the legal, ethical, and procedural terrain is unrecognisable. When a machine makes a decision that leads to harm, *who is at fault*? And perhaps more challengingly, *what does fault even mean* in a distributed system where decisions are made not in the moment, but in the design stage, the coding process, and the algorithmic tuning that may have occurred months prior?

The legal concept of fault is predicated on the idea of agency—that someone, somewhere, chose a course of action that breached a duty of care. But in autonomous systems, action is diffused through layers of abstraction. A braking decision, for example, might be the result of a sensor reading, interpreted by a perception model, cross-referenced against a probabilistic risk matrix, and then executed by an actuator under firmware control. Each component functions correctly within its domain, but the system as a whole might still produce a harmful outcome. There is no malice, no recklessness, no fatigue. There is only a failure in prediction. And in such cases, the traditional tools of tort law—intent, negligence, foreseeability—begin to buckle under the complexity of machine-led causation.

A well-documented example is the 2018 fatal crash in Tempe, Arizona, involving a pedestrian, Elaine Herzberg, and a self-driving Volvo SUV being tested by Uber. The onboard software identified Herzberg but failed to classify her as a pedestrian in time to react. A backup safety driver was present but distracted, watching a video on her phone. The vehicle did not brake. The result was fatal. Prosecutors eventually charged the safety driver with negligent homicide, but not Uber. The decision drew criticism from legal scholars who argued that the system's design flaws—specifically, its inability to anticipate common pedestrian behaviour—were more structurally significant than a lapse in driver vigilance. The ambiguity laid bare a new reality: *it is possible for a machine to cause death without a clear culprit to charge.*

This problem is not hypothetical. As AVs become more prevalent, so too will incidents where machines act in ways that, while procedurally correct by system standards, result in harm that feels intuitively blameworthy. The public response to such events often demands accountability, but accountability without agency quickly devolves into confusion or scapegoating. Blame is redirected toward the most proximate human—the safety driver, the vehicle operator, the pedestrian who crossed unexpectedly. Yet these individuals may have little or no influence over the decisions that led to the incident. Meanwhile, those with real influence—software architects, corporate product leads, AI ethics teams—operate at a remove that shields them from legal consequence.

Attempts to retrofit existing legal categories to AV incidents have so far yielded inconsistent outcomes. In some jurisdictions, product liability law is being tested as a mechanism for assigning fault to manufacturers or software providers. This approach treats the AV as a consumer product and seeks to determine whether it was "defective" in design or execution. But the challenge lies in defining *defectiveness* in a system where failure modes are probabilistic rather than deterministic. A human driver who fails to brake is clearly at fault. A system that fails to predict an edge case—a cyclist swerving to avoid debris, a dog darting into the road—is not clearly "defective" if it performed according to its parameters. It simply encountered an unmodelled scenario. In such cases, the language of defect collapses under the weight of statistical ambiguity.

Moreover, software liability raises questions of transparency and disclosure. Many AV systems operate as black boxes—proprietary, opaque, and legally protected by intellectual property law. When an incident occurs, investigators may be unable to access the relevant data or decipher the decision logic. Without transparency, fault becomes not only hard to assign but hard to investigate. This is compounded by the fact that AV software evolves constantly. Over-the-air updates can change system behaviour weekly or even daily. An update pushed the night before a crash may alter braking thresholds, pedestrian detection sensitivity, or route selection priorities. In such an environment, *the evidentiary trail is ephemeral*, scattered across cloud servers, dev logs, and version histories.

Even where fault can be identified, the remedy may be inadequate. Traditional insurance models depend on insuring a person—someone with a licence, a driving history, and a premium calibrated to individual risk. But in an AV world, it is not the driver but the system that determines outcomes. This has led to calls for shifting insurance responsibility to manufacturers or mobility service providers. Volvo, for instance, has pledged to accept liability for its autonomous systems when operating in fully autonomous mode. But such commitments are rare, and the legal framework to support them is still nascent. What happens when multiple companies are involved—the vehicle manufacturer, the software supplier, the mapping provider, the fleet operator? *Distributed design leads to distributed liability*, and distributed liability often leads to no accountability at all.

The ethical implications are equally complex. In human-operated vehicles, moral responsibility is relatively clear: if a driver chooses to speed, drink, or text while driving, they bear the consequences. With AVs, responsibility becomes a function of design choices made by people who never appear at the scene. Who decided that the system should prioritise passenger safety over pedestrian safety in no-win scenarios? Who calibrated the reaction threshold for hard braking to minimise false positives but increased stopping distance? These are *moral decisions made in code*, yet there is no moral agent visibly attached to them. The software team is often anonymous, and the decisions themselves may be emergent properties of machine learning rather than explicitly programmed outcomes.

The tension between technical inevitability and moral culpability creates a zone of ethical diffusion. Harm is real, but accountability is dispersed. This is particularly troubling in cases where systemic bias enters through data. If an AV system performs poorly in recognising dark-skinned pedestrians or disproportionately avoids certain neighbourhoods due to data-driven risk models, the resulting harm is structural, not incidental. Yet the law, as currently structured, is ill-equipped to address harm without intention. *Discrimination by design is harder to litigate than discrimination by decision.*

Some legal scholars have proposed creating new categories of responsibility tailored to AV systems—"algorithmic duty of care," for example, or "systemic negligence." These efforts are intellectually promising but legally untested. Courts are conservative by nature and tend to adapt slowly to technological disruption. Meanwhile, the deployment of AVs accelerates. Vehicles are already operating on public roads, interacting with human drivers, cyclists, and pedestrians in complex environments. The lag between deployment and legal adaptation creates a regulatory vacuum—one in which *harm can occur without remedy.*

The political implications of this vacuum are considerable. As citizens, we expect that when harm occurs, the system will respond—not only to compensate victims, but to prevent recurrence. When that response fails—or worse, appears absent—the legitimacy of the system erodes. Public trust in AVs is not only about technical reliability; it is about the assurance that *someone is responsible.* If that assurance evaporates, so too does the social licence to operate.

What is required, then, is a reconceptualisation of responsibility that accounts for *the complexity and distributed nature of AV systems.* This does not mean abandoning the principle of accountability. On the contrary, it means extending it—recognising that responsibility must be engineered into the system as rigorously as safety or efficiency. It means demanding transparency as a condition of deployment, not as a post-incident courtesy. It means developing legal frameworks that can accommodate systemic causation, not just individual fault. And it means building a culture of ethical design, in which those who shape machine behaviour understand that they are shaping the conditions of human safety, dignity, and justice.

Responsibility, in the age of autonomous vehicles, is not something that will reassert itself automatically. It must be defined, distributed, and defended in new ways. Otherwise, as the wheel recedes into metaphor and the human slips from the driver's seat to the passenger's, we may discover too late that accountability has followed suit—disappearing quietly into the rearview, just when we needed it most.

The Ownership Question

Ownership, once a cornerstone of mobility and personal autonomy, is quietly being disassembled. For more than a century, the privately owned car symbolised not only the freedom to move but also a form of self-determination. The driver chose when to leave, which route to take, whether to detour or to park and walk. Vehicle ownership offered control, privacy, and identity—status and function rolled into a single machine. Yet as the automotive industry shifts toward service-based models—platforms over products, subscriptions over possessions—the entire meaning of vehicle control and personal mobility is undergoing a structural realignment. The implications are not simply economic or logistical; they are political, cultural, and deeply personal.

The rise of the autonomous vehicle accelerates this transformation. A privately owned autonomous car is technically possible, but operationally inefficient. AVs gain their greatest economic advantage when they are deployed in shared fleets—circulating continuously, picking up new passengers, optimising routing based on real-time data. The logic of automation dovetails seamlessly with the logic of platform capitalism. Companies are not building autonomous cars for people to own; they are building them for people to *access* on demand. In this model, vehicles are nodes in a service network, not individual assets. Ownership is reconfigured as usage, and usage is managed through digital platforms governed by terms of service—not by rights in the traditional sense, but by permissions that can be granted, limited, or revoked.

This shift has been marketed as progress. And on the surface, it is appealing: no car payments, no insurance worries, no maintenance responsibilities, no need for parking. The appeal is especially strong in urban areas where the burdens of ownership often outweigh the benefits. Yet the reclassification of mobility as a service, rather than a right or a resource, alters the balance of power between individual and system. When one owns a vehicle, one has a degree of autonomy that resists systemic interference. The car can be used as one sees fit, within legal limits. When one subscribes to a mobility platform, every action is subject to conditions: where you can go, when you can travel, how much it will cost, and whether the system deems your request viable. *What is gained in convenience is lost in sovereignty.*

Consider the example of geofencing—technological perimeters that define where AVs can operate or serve passengers. These zones are not neutral. They are shaped by business models, local regulation, and risk management strategies. A platform may choose not to serve certain areas due to low profitability, perceived danger, or infrastructure incompatibility. The result is a stratified transport map in which some neighbourhoods enjoy frequent, low-cost, high-speed access, while others remain isolated. The user cannot challenge this allocation. They cannot reconfigure the map or demand access as a right. They can only wait for the platform to decide that expansion is economically justified.

In a system of private ownership, one could at least choose to drive into such areas oneself. In a service-based model, one must request permission—often algorithmically mediated and commercially filtered. The transformation is not trivial. *It recasts the citizen as a consumer and the road as a product.*

This has consequences for data autonomy as well. Owning a car generates data, but that data remains largely under the driver's control. In contrast, participating in a mobility platform involves the continuous extraction of behavioural data: location history, time of travel, frequency of trips, route preferences, co-passengers, and even biometric identifiers in some systems. This data is owned and processed by the platform, not the individual. It becomes part of the commercial logic of optimisation—used to improve efficiency, yes, but also to shape pricing, access, and even user ranking.

In theory, one can choose not to participate. In practice, opting out of platform-based transport increasingly resembles opting out of urban life itself. As more services—from deliveries to employment—depend on seamless mobility, refusing the conditions of access may impose social or economic penalties. This is not a traditional coercion, but a soft compulsion—a narrowing of alternatives until the service becomes indispensable. And once indispensable, it ceases to be optional. The absence of ownership thus becomes the presence of dependency.

There are broader historical patterns worth recalling here. The 20th century saw a deliberate push toward car ownership, often at the expense of public transport and walkable city planning. The result was a spatial and social architecture built around individual control, but also around exclusion, congestion, and inequality. Today, the pendulum appears to be swinging back—not toward shared public infrastructure, but toward privately operated shared platforms. The difference is crucial. Public transport, however flawed, is governed by public mandates. It is accountable through elections, oversight, and civic engagement. Mobility-as-a-service platforms are accountable to shareholders and operational metrics. They operate on terms designed to optimise return, not equity.

Some argue that this is a price worth paying—that we should welcome the end of private vehicle ownership as an environmental and social good. Fewer cars on the road, more efficient routing, reduced emissions, and less urban space devoted to parking are undeniably worthwhile goals. But these goals do not require the elimination of ownership; they require the elimination of *wasteful* ownership. The binary framing—own or share, control or delegate—is a distraction. The real question is how to retain meaningful autonomy within shared systems. Can users set terms? Can they demand transparency? Can they challenge platform decisions or appeal system exclusions? These questions are rarely addressed in the promotional material of AV companies, but they lie at the heart of what it means to navigate the world freely.

There are also legal implications. Ownership has traditionally carried with it a bundle of rights: the right to repair, the right to customise, the right to use or not use at one's discretion. As cars become increasingly software-driven and service-linked, these rights are being quietly eroded. A user may not be allowed to modify an AV's behaviour—even if for legitimate or personal reasons. They may be required to accept updates that change system functionality. In extreme cases, vehicles may be remotely disabled for contract violations or payment lapses. What was once a matter of private use becomes a matter of platform compliance. *The user remains in the car, but the car no longer belongs to them—in practice or in principle.*

This trend is not limited to AVs. Similar patterns have emerged in digital media, home automation, and even agriculture, where farmers lease "smart tractors" that prevent them from making unauthorised repairs. The logic is always the same: sell the experience, retain the asset, and lock the user into a system of dependence masquerading as convenience. The consequences for autonomy are cumulative and profound.

To resist this trend does not mean clinging to outdated models of ownership. Rather, it means reimagining ownership for a world of intelligent systems. *What would it mean to own an AV in a way that preserved agency while enabling the benefits of networked transport?* What kind of policies, standards, or cooperatives would be needed to make shared ownership viable? Could public institutions play a role in ensuring that service models serve public interest—not just platform growth?

Without such questions, the shift from ownership to service risks becoming irreversible not by design but by inertia. A generation that never experiences ownership may come to see autonomy as antiquated, unnecessary, even irresponsible. And in that shift, we lose not just a legal category, but a cultural anchor—a sense that mobility is something we *do*, not something that is done *to* us.

The future of transport is not about who drives. It is about who decides. In the age of the autonomous vehicle, *owning nothing and controlling nothing are not the same thing—but they are becoming dangerously close.* To reclaim autonomy in this context means demanding not just access, but authorship—over the systems, the standards, and the rules that govern how, where, and why we move. The ownership question is, in the end, a question of power. And it is one we cannot afford to leave unanswered.

Chapter 3
The Ethics Engine

When a child dashes into the street, should an autonomous vehicle swerve into an oncoming lane, endangering its passengers, or brake abruptly, risking the child? Should it treat a pedestrian differently from a cyclist, an elderly person differently from a teenager, a rule-follower differently from a jaywalker? These are not fanciful hypotheticals; they are pressing programming dilemmas debated in boardrooms and research labs. At their core lies an uncomfortable truth: the forthcoming revolution in transport is not merely technical—it is profoundly moral. Autonomous vehicles are now asked to make choices once reserved for human judgement, turning ethics from a matter of conscience into lines of code.

The notion that machines can—or should—resolve moral dilemmas is deeply contested. Some maintain that ethical encoding is necessary to reduce harm in extreme scenarios; others insist the endeavour is intrinsically flawed, since ethics resist universal formulas. Yet the discussion has left the abstract and entered the tangible: AVs on public roads make real-time decisions governed not by intuition but by programmed rules, filtered through probabilistic models, constrained by sensor limitations, network lag, and computational logic. Ethics, once philosophical, has become operational.

But whose ethics guide these machines? Whose lives are accounted for, and by what criteria? Engineers, rarely trained in moral philosophy, are now shaping the very frameworks of life-and-death decisions. A German vehicle may reflect national regulatory priorities, favouring adherence and efficiency. An American platform might prioritise passenger protection or legal defensibility. A Chinese system may encode state objectives emphasising social order. None is inherently superior, yet each embeds assumptions with profound consequences: code that executes in milliseconds carries ethical weight in real time.

What emerges is a new layer of "moral infrastructure," largely invisible until it fails. Like roads, bridges, or power grids, it reflects the interests of those who design and fund it. Compounding the issue is opacity: AV firms, invoking intellectual property, are under no obligation to reveal the logic governing their ethical choices. The result is a troubling asymmetry: machines make ethical decisions in public, while citizens have no practical means to examine or contest them. Transparency, once the bedrock of accountability, is rendered optional.

Consider the digital iteration of the classic trolley problem. Engineers often dismiss it as unrealistic, yet it exposes the central tension: AVs are tasked with harm reduction, but harm is unevenly distributed. Who defines it? How are trade-offs made between probability and consequence, between individual welfare and collective risk? Machine learning systems resolve such questions statistically, but the consequences affect living people. A model designed to minimise overall fatalities may consistently endanger particular groups—urban residents, the mobility-impaired, or those whose behaviour is unpredictable. Ethical programming, then, can become a vehicle for systemic inequity.

Bias, moreover, is rarely absent. AVs learn from vast datasets reflecting decades of human prejudice in traffic enforcement, infrastructure, and driver behaviour. Without correction, these biases are not merely replicated—they are amplified. A human driver's partiality affects one choice; an algorithm propagates it across thousands of decisions daily. This is not simply a technical problem but a societal and moral one: mechanised bias operates at scale, anonymised yet potent.

Compounding the challenge is the lack of consensus. Guidelines exist, but no universal ethical framework governs AV conduct. Should passenger safety take precedence? Should collective risk override individual protection? Should legal liability outweigh moral calculation? Location, context, or user profile—should they influence the machine's choice? These questions remain unresolved, forcing machines to enact judgments that society itself has not agreed upon.

Yet the market presses on. Companies face immense pressure to launch, scale, and monetise technology. Ethical deliberation is costly in time, capital, and regulatory exposure. Safer, more transparent systems may be slower, more expensive, or less profitable. Many firms invest sincerely in safety and maintain ethics boards, yet absent enforceable standards, accountability remains voluntary and uneven. Ethics, in practice, is subordinated to expediency.

This raises deeper questions of responsibility. Can a machine be accountable for decisions it was programmed to make? Should liability fall on developers, operators, data providers, or end users? Legal frameworks lag behind, leaving responsibility diffused across supply chains and software layers. Without clarity, accountability erodes—and when no one is answerable, everyone is vulnerable.

Public trust in AVs hinges not solely on safety, but on moral coherence. People tolerate occasional malfunctions; they will not tolerate choices that appear unethical. A vehicle sacrificing a pedestrian to save passengers will provoke outrage, even if statistically justified. Acceptance depends not only on outcomes but on the perceived fairness, transparency, and governability of decision-making.

This is the crucial point: autonomous vehicles must be legible in their ethics. Safety alone is insufficient. Decisions of moral weight cannot remain hidden behind proprietary code. They must be visible, contestable, and shaped through democratic oversight. Otherwise, we risk a future where moral decisions are rendered silently, invisibly, and without accountability—not by human judgement, but in spite of it.

In this emerging landscape, the ethical engine is the true heart of the AV system—not sensors, neural networks, or drive trains. It decides who is protected, who is endangered, and how justice is operationalised on the streets. Like any engine, it must be constructed with care: not merely for performance, but for principle.

The Moral Logic of Machines

A car that decides who lives and who dies is no longer a thought experiment—it is a product under development. As autonomous vehicles transition from test tracks to public streets, they encounter more than just traffic lights and pedestrians; they enter the domain of ethical decision-making, where machine logic must grapple with human complexity. In a high-speed scenario, if collision is unavoidable, should the vehicle prioritise the lives of its passengers or of pedestrians? Should it consider age, behaviour, or fault? What if the passengers include children and the pedestrians are elderly? These are not rhetorical questions. Engineers are being asked to translate them into algorithms, and with that translation comes an epochal shift: *the delegation of moral agency to machines.*

Unlike traditional driver safety systems, which serve to augment human judgement—like anti-lock brakes or lane assist—autonomous vehicles displace it. They do not merely advise; they decide. And with that, the very structure of decision-making changes. A human driver might act from instinct, emotion, or split-second ethical reasoning. A machine acts from predefined logic trees and probabilistic modelling. It is not that AVs make no moral choices, but that *their morality is pre-engineered,* locked into code well before a scenario arises. Every decision made by an autonomous vehicle is a delayed consequence of a series of human choices: which data to train on, which risks to prioritise, which outcomes to consider acceptable.

The most well-known articulation of this problem is a modern twist on the classic "trolley dilemma." Suppose an AV must choose between swerving into a barrier and killing its passenger, or continuing forward and hitting three jaywalking pedestrians. The calculation seems straightforward: minimise loss of life. Yet as soon as variables are introduced—children versus adults, lawful behaviour versus recklessness, socioeconomic factors, even racial demographics—the situation becomes ethically volatile. It is no longer a question of utility, but of *values.* Should an AV ever make a choice based on probabilistic estimates of who is more likely to survive? Is it ethical for a machine to weigh one life over another, based on anything other than number?

Some developers sidestep the problem by programming AVs to follow traffic laws absolutely, deferring to rules rather than judgement. This may seem like a clean solution—no ethical discretion, only legality. But laws are not comprehensive moral systems. They contain gaps, contradictions, and contextual blind spots. A legalistic AV might brake for a red light even if doing so would cause a high-speed rear-end collision. It might avoid jaywalkers but ignore the needs of cyclists swerving to avoid debris. *Lawful behaviour does not equate to moral behaviour*—and in some cases, it may actively prevent ethical outcomes.

Even when engineers attempt to "build in" ethics using frameworks like utilitarianism, deontology, or virtue ethics, the translation into code is incomplete. These systems rely on qualitative judgement, which algorithms struggle to replicate. A utilitarian AV may be designed to reduce overall harm, but how does it weigh the potential suffering of survivors? A deontological AV may follow strict rules—do not kill, do not harm—but what if every available action causes harm? In the world of AVs, ethical philosophy must become machine-readable. And that requires abstraction, simplification, and ultimately, *moral reductionism.*

Consider a real-world example: in 2018, an Uber self-driving test vehicle in Arizona fatally struck a pedestrian. The onboard system had identified the individual but classified her erratically as an unknown object, a vehicle, and then a bicycle. The emergency braking system was disabled, and the safety driver was distracted. This incident did not result from a single programming failure, but from a cascade of ethical and technical assumptions—about classification thresholds, system overrides, and the trade-offs between false positives and missed detections. The tragedy was not just a software bug; it was a failure of moral architecture.

Moreover, AVs must resolve ethical dilemmas not in isolation, but within ecosystems of unpredictability. Road users include not only drivers but cyclists, pedestrians, animals, construction workers, and people with disabilities—all exhibiting diverse behaviours that challenge uniform decision-making. A human might make an eye-contact judgement or gesture acknowledgement. A machine must calculate intent from pattern recognition, often with incomplete data. The absence of intuition, emotion, and context-awareness is not a limitation that can be patched—it is a defining condition of machine reasoning. *And when moral logic meets real-world ambiguity, machines are forced to decide anyway.*

One might argue that moral inconsistencies already exist in human driving, and that AVs, at scale, will statistically reduce harm even if they cannot handle edge cases perfectly. That is likely true. But the issue is not whether AVs can outperform humans in aggregate—it is whether the way they fail is acceptable to the societies they serve. If a human driver makes a tragic error, we tend to treat it as a misfortune, even when it results in loss of life. If a machine makes the same error, especially one rooted in code or data, it is seen as a systemic failure—a design flaw, not a personal lapse. *The threshold for forgiveness is lower when the agent is human and higher when the agent is synthetic.*

This asymmetry becomes even sharper when one considers who controls the moral architecture. AVs are built by private companies—many of them tech giants with global reach. These firms are not democratically elected, nor are they transparent about their ethical frameworks. As a result, decisions about who is protected, who is prioritised, and what risks are acceptable are increasingly made in corporate conference rooms rather than public forums. Consumers have no clear way to know whether their vehicle would prioritise their life or someone else's in a crisis. Regulators, lacking the technical access or legislative mandate, struggle to keep pace. *We are not just outsourcing driving—we are outsourcing decision rights.*

Efforts to embed public values into AV systems have been uneven. The "Moral Machine" experiment, conducted by MIT, collected millions of responses globally to determine public preferences in AV decision-making scenarios. The results revealed vast differences by region, culture, and socioeconomic context. Some societies valued youth over age, others prioritised lawful behaviour. In some regions, animals were given moral standing; in others, they were not. The key finding was not consensus—it was divergence. *There is no universal ethical formula that satisfies all cultural contexts,* yet AV systems must be programmed with a single set of rules.

This divergence matters because AVs do not operate in vacuums—they navigate within the deeply social, symbolic space of the road. Every vehicle on a street reflects assumptions about what is acceptable, what is negotiable, and what is off-limits. When those assumptions are encoded into software, they cease to be fluid and become fixed. Ethical flexibility, once the domain of human judgement, becomes a liability in code. And yet the absence of flexibility creates its own dangers: *rigid systems cannot adapt to novel moral contexts*, and thus may act predictably—but in ways that feel deeply inhuman.

One response to this impasse has been the call for "value-sensitive design"—an approach that integrates ethical considerations from the beginning of the engineering process. Rather than retrofitting moral logic after the fact, value-sensitive design seeks to embed pluralistic perspectives into the lifecycle of product development. But this approach demands more than technical fluency. It requires interdisciplinary collaboration, public engagement, and a willingness to surface conflicts that may slow down deployment. In an industry driven by speed-to-market, that is a difficult proposition.

Ultimately, the challenge of programming moral logic into machines is not just a software problem—it is a societal one. It forces us to confront the limitations of our own moral systems, our capacity for consensus, and our willingness to trust decision-making to entities that neither feel pain nor carry legal personhood. The AV dilemma is not whether machines will act ethically—it is whether *we are willing to live with the consequences* of how we define ethics in code.

As this transformation unfolds, moral decision-making will become one of the most consequential components of AV architecture. No longer a matter for philosophy seminars, ethics will be built into chips, tested in simulations, deployed across fleets, and evaluated through statistics. And yet each statistic will contain a story—of who was saved, who was harmed, and who made that choice long before the wheels ever turned. The moral logic of machines, once a theoretical curiosity, has become a foundational issue for the future of movement, agency, and justice in a world that is being driven by code.

Whose Values Get Programmed?

In 2018, a research study known as the Moral Machine experiment surveyed over two million people across 233 countries and territories, presenting them with a series of ethical dilemmas a self-driving car might face—such as whether to swerve and kill one pedestrian or stay the course and hit five. The responses revealed striking geographical patterns: respondents in Western countries tended to prioritise younger over older lives, while some East Asian participants were more inclined to preserve the group over the individual. The study laid bare a truth that the AV industry has largely glossed over—there is no universal moral compass to program into machines. The ethical logic of autonomous vehicles is not a neutral algorithm but an encoded worldview, and whose worldview it reflects is a question with vast political, cultural, and commercial consequences.

At the heart of the AV revolution is a conceit of objectivity—that technology, especially one so heavily dependent on data and code, will be impartial, rational, and fair. But this premise collapses under scrutiny. A vehicle's decision-making process in critical moments is not merely a matter of physics or sensor readings. It is shaped by design choices made by engineers, policy directives from governments, and risk tolerances from insurers. Each of these stakeholders carries assumptions about what constitutes acceptable harm, who should bear risk, and what societal values take precedence when trade-offs are inevitable. When a pedestrian darts into the road and the AV must choose between swerving into a tree or colliding with the person, its response is not just a technical matter—it is an ethical act, predetermined by layers of programmed logic.

This logic, however, is not generated in a vacuum. In the United States, where AV development has been largely driven by private companies such as Waymo, Tesla, and Cruise, the dominant value set is one of individual freedom, market optimisation, and product liability avoidance. Decisions about how vehicles behave in edge-case scenarios often reflect legal defensibility more than ethical deliberation. Contrast this with Germany, where the Federal Ministry of Transport convened an Ethics Commission in 2017 to establish AV programming guidelines that explicitly prohibit decisions based on personal characteristics such as age, gender, or race. The result is a codified approach to AV ethics shaped by a post-war German commitment to human dignity and a deep wariness of algorithmic discrimination.

China's trajectory introduces a third paradigm—state-anchored technological development with collectivist values informing system priorities. Here, AVs are seen not only as consumer goods but as instruments of national infrastructure, public order, and social engineering. Ethical decisions in vehicle software may lean toward minimising systemic disruption rather than privileging individual lives. Moreover, AV data in China is heavily integrated into broader state-run surveillance systems, underscoring a philosophical difference about where the boundaries of technology and governance lie.

These national variations are not academic exercises—they translate into real divergences in how AVs may behave, whom they may protect or endanger, and which moral instincts they enshrine. A pedestrian crossing a street in Stuttgart may encounter a car coded to make decisions based on neutrality of personhood. That same pedestrian in Phoenix may face a vehicle optimised to minimise corporate liability. In Shenzhen, it may be the broader utility of traffic efficiency that determines whether the brakes are hit or the trajectory maintained. In this context, *ethical localisation* becomes a new challenge in software development: should vehicles behave differently based on cultural context? And if so, how does one define cultural context within a system of rigid logic?

The notion that values can be modularised and localised within software opens another thorny question—who decides which values apply where? At present, AV companies make these decisions largely in-house, with proprietary algorithms shielded under intellectual property law. Regulatory oversight remains fragmented and uneven, with most governments still struggling to establish coherent frameworks for AV deployment, let alone intervene in their ethical core. This leaves private firms in the extraordinary position of acting as de facto moral legislators, encoding ethical principles into systems with life-and-death authority, but without democratic input or public transparency.

This opacity is particularly troubling given the financial stakes. Ethics is not just a philosophical consideration; it is increasingly a commercial differentiator. Firms that can market their vehicles as "safest" or "most reliable" gain a competitive advantage, yet there is no agreed-upon benchmark for what that safety entails ethically. One company may optimise for protecting passengers at all costs, while another balances harm more broadly across all road users. Without standardised guidelines or public accountability, these choices risk being driven less by ethical reasoning and more by branding strategy, shareholder pressure, or geopolitical incentives.

Further complicating the terrain is the global supply chain of AV systems. A vehicle assembled in Mexico might carry LiDAR systems developed in Israel, decision-making software coded in California, and ethical rule-sets derived from European white papers. Each layer introduces a different cultural and philosophical input, often without coherent integration. The result can be patchwork morality—internally consistent within subsystems but incoherent when unified under real-world pressure. This introduces not only legal uncertainty but the possibility of machine behaviours that are culturally tone-deaf or ethically indefensible in particular contexts.

Philosophers and ethicists have long debated the possibility of universal ethics. Autonomous vehicles force this debate out of the seminar room and onto public roads. They raise the stakes: whose life matters most in a no-win scenario? Should an AV prioritise the safety of its occupant over a jaywalking child? What role do cultural expectations, historical injustices, or demographic patterns play in designing answers to those questions? These are no longer speculative musings—they are live engineering decisions being made every day.

Yet public participation in these ethical formulations remains minimal. Most people assume AVs will "do the right thing," without realising that *the right thing* is itself a coded decision that varies depending on where the car is built, who funded its design, and what the software team deemed acceptable trade-offs. The absence of a public mandate means that ethical frameworks may reflect narrow institutional logics rather than broader societal deliberation. This asymmetry between who builds the system and who bears its consequences is a democratic failure in waiting.

To rectify this, some scholars advocate for open-source ethics—a radical transparency in how decision-making logic is coded, debated, and tested. Others push for national or international regulatory bodies to establish enforceable ethical standards, akin to medical bioethics or aviation safety protocols. There is also growing interest in participatory design models that invite community input into system priorities. These proposals share a common aim: to ensure that the moral logic of machines is not dictated solely by market actors but is shaped by collective human reflection.

Still, even with greater participation and oversight, the core challenge remains: ethics does not scale neatly. What is just in one context may be unjust in another. Efforts to build consensus often collapse into either bland generalities or irreconcilable conflict. Yet the very attempt to confront these complexities—rather than bury them beneath a sheen of technological inevitability—is what responsible innovation demands. It is not enough to build vehicles that can see, brake, and navigate. They must also *discern*, *prioritise*, and *justify*. And in doing so, they must carry forward the values of the societies that unleash them—not invisibly, but intentionally, accountably, and above all, humanely.

Risk, Bias, and the Politics of Safety

One of the least acknowledged yet most consequential dimensions of autonomous vehicle (AV) development lies in how risk is computed, distributed, and ultimately absorbed. AVs are often presented as the culmination of safety-driven innovation—heralded as the technology that will end traffic fatalities, eliminate human error, and deliver a transport utopia. But this narrative eclipses a more complex and discomforting reality: the very algorithms designed to make our roads safer may also be embedding new forms of inequity and bias, silently redistributing risk across demographic, geographic, and socio-economic lines. Safety, in this context, is not an absolute—it is a selective outcome shaped by design choices, data sets, and strategic compromises. *Autonomy, it turns out, does not float free from politics.*

To understand this, we must begin with a simple but sobering truth: *machines do not make ethical decisions—they execute code written by people who do.* And people, for all their brilliance and intention, bring to the table their own assumptions, priorities, and blind spots. When an AV is forced to "choose" between swerving into a pedestrian or protecting its passenger, that is not the vehicle making a decision—it is a logic tree predetermined by design. And that logic, in turn, is shaped by countless factors: the company's safety philosophy, regional regulations, legal liability considerations, and even cultural norms. These codified decisions, especially when not fully transparent, reflect *power-laden judgments* about whose life is valued more, which types of accidents are "acceptable," and which populations are statistically worth exposing to greater harm.

A 2020 study from the Georgia Institute of Technology and MIT Media Lab revealed that many AV training datasets struggle to accurately detect individuals with darker skin tones, particularly under low-light conditions. This is not merely a technical shortcoming; it's a moral failing of scope and representation. If your data corpus underrepresents certain communities, your system will be less safe for them. And when your sensors struggle with night-time detection in urban areas—often home to marginalised populations—then the "smartest" car on the road may pose an outsized risk to those who are already structurally disadvantaged. *Bias in, risk out.*

Moreover, safety models are rarely static. As AVs learn through continual data collection and real-time machine learning, they adapt based on environmental feedback. But the environments themselves are unequal. Streetscapes vary not only in road conditions or signage, but in policing patterns, pedestrian behaviour, and infrastructure investment. AVs trained primarily in wealthy, well-manicured suburbs of California may exhibit very different behaviours when navigating through the dense, erratic, and underfunded roadways of, say, Baltimore or Mumbai. In that sense, algorithmic safety is *location-contingent*—designed for, and often optimised by, the contexts in which testing and validation occur. A car that performs safely in Palo Alto may behave unpredictably in Johannesburg. Yet the safety claims issued by manufacturers rarely include these geographic caveats.

Another layer of complexity arises from the way risk is communicated, particularly to regulators and the public. Most AV companies tout their "miles without incident" or "disengagement rates"—metrics that are vague, unaudited, and often misleading. These figures provide a sense of technical performance but obscure critical ethical considerations. What kind of incidents are excluded? Were safety drivers forced to take control due to sensor confusion in minority neighbourhoods? Were there near-misses involving cyclists or children? *What we measure shapes what we see, and what we ignore silently grows.*

Consider also the asymmetry of legal liability. When a human driver errs, the law has clear mechanisms for assigning fault. But when a self-driving system fails, the trail of accountability becomes murky. Is the software engineer responsible for a misclassified pedestrian? Is it the data scientist who built the training set? The executive who signed off on deployment? Or the insurance company underwriting fleet operations? This ambiguity not only complicates compensation and justice—it *incentivises opacity*. The more diffused the fault, the harder it is to pin down institutional responsibility. This blurring of legal lines effectively distributes risk downward—onto users, victims, and local governments—while protecting the platforms that profit.

The implications extend further. Many AV companies are shifting towards fleet models—ride-hailing services where cars are operated as shared, on-demand assets. This setup introduces corporate incentives to optimise for speed, throughput, and cost reduction. Safety, in such a context, becomes one of many competing variables. And when margin pressures grow, risk calibration may skew towards prioritising the system's overall efficiency rather than *each individual's safety*. In other words, decisions may be made at the population level, not the person level. It is not hard to imagine an AV system choosing a route that marginally increases collision risk in a low-income neighbourhood because it trims two minutes off a high-volume corridor's average commute time. *The politics of safety are encoded in the economics of scale.*

Furthermore, the ethical expectations of safety differ across cultures. In Germany, traffic laws emphasise collective responsibility and caution, while in the United States, the legal system is more adversarial and litigious. What constitutes a "reasonable" risk in Beijing may be unacceptable in Helsinki. As AV companies expand globally, they are faced with programming decisions that do not map neatly across jurisdictions. The infamous "trolley problem" may be a Western philosophical abstraction, but in real-world AV design, some version of it is unavoidable: who gets protected, who gets exposed, and who gets forgotten? Without regulatory harmonisation or meaningful public discourse, these decisions are often made privately, within corporate labs and boardrooms. *And that is where ethical drift begins.*

Crucially, the narrative of risk tends to erase who has the power to speak about safety, and who gets left out. Marginalised communities, historically subjected to over-policing and under-protection, now face a new form of invisible governance—algorithmic surveillance on wheels. AVs rely heavily on computer vision and location tracking. In doing so, they introduce a rolling data apparatus that continuously monitors its environment. The data captured—faces, movements, street activity—is rarely subjected to meaningful oversight. Thus, safety becomes entwined with surveillance. In the name of protection, a new layer of data extraction emerges. For those already wary of institutional trust, this does not feel like innovation—it feels like escalation.

None of this is to say that AVs are inherently unsafe or malicious. Quite the opposite: they hold extraordinary potential to reduce traffic deaths, assist the disabled, and lower barriers to mobility. But to realise that promise equitably, we must interrogate how and where risk is externalised, and who gets to participate in deciding what "safe enough" actually means. *Safety is not just a metric—it is a value judgement.* One that should not be left to engineers alone, nor deferred to regulatory agencies after deployment.

True safety in the AV era will demand new forms of transparency, participatory design, and democratic oversight. It will require companies to open their datasets, explain their models, and subject their algorithms to ethical audit—not just technical verification. It will require regulators to expand their definition of safety beyond collision statistics, to include fairness, consent, and social trust. And it will require the public to remain vigilant—to understand that while the car may be self-driving, the risks are still collectively borne.

Transparency vs. Trade Secrets

When a machine makes life-and-death decisions, the public tends to ask a simple question: how, exactly, was that decision made? With autonomous vehicles (AVs), however, the answer is anything but simple. What should be open knowledge in a public safety context is often cloaked in proprietary secrecy. The algorithms guiding AVs operate not in a neutral vacuum but within intricate layers of commercial strategy, intellectual property claims, and regulatory ambiguity. At the heart of the debate lies a hard truth: *the very logic that governs whether a self-driving car brakes, swerves, or accelerates is shielded from scrutiny by design.*

The AV industry—led by a small cadre of tech giants and mobility startups—has fiercely guarded its software as a competitive asset. This is understandable from a business standpoint. Algorithms represent years of R&D, vast investments, and hard-won intellectual edge. Disclosing them in full would effectively surrender competitive advantage. Yet, unlike a recommendation engine or ad-targeting model, the outputs of AV decision-making can result in physical harm or death. In such cases, corporate secrecy can obstruct justice, stall regulation, and undermine public trust. *The very opacity that protects innovation can also shield negligence, bias, or design flaws.*

This tension came into sharp relief following the 2018 fatality involving an Uber AV in Tempe, Arizona. Investigations revealed not only technical failures—such as the vehicle's failure to classify a pedestrian properly—but also design decisions related to how aggressively the system handled uncertain detections. Yet Uber's full decision-making logic remained sealed from the public, citing trade secrets. Regulators received redacted documentation. The press was handed partial narratives. The result was a fragmented understanding of what had gone wrong—and an inability for the wider public or peer engineers to scrutinise, learn, or improve upon the flaw.

Contrast this with aviation, where safety incidents are subject to thorough, often public, investigation. Black box data is recoverable, and systemic reviews typically follow. Manufacturers, operators, and regulators participate in a shared safety culture. AVs, despite operating in similarly high-stakes environments, are not held to that same standard. There is no equivalent of the Federal Aviation Administration or the National Transportation Safety Board for self-driving software logic. *Instead, AV companies effectively self-investigate while regulators look on, underfunded and often ill-equipped to interrogate algorithmic logic.*

This asymmetry is exacerbated by the complexity of machine learning itself. Unlike hand-coded systems, AVs often rely on deep learning models whose internal decision chains are notoriously difficult to interpret—even by their creators. "Black box" doesn't just refer to corporate secrecy; it also reflects the inherent opacity of the models in use. As a result, when incidents occur, distinguishing between a company's unwillingness to disclose and its technical inability to explain becomes murky. *The public is left to accept that machines are operating in their midst, guided by logic neither transparent nor intelligible.*

One proposed remedy is algorithmic auditability: requiring AV firms to log, store, and disclose the rationale behind key decisions made in critical contexts—particularly in edge cases involving injury or death. Such logs wouldn't necessarily reveal source code or commercial secrets but would provide regulators and the public with a window into the vehicle's reasoning. Despite growing support for such measures among ethicists and legal scholars, industry resistance remains strong. Some firms argue that algorithmic logs can still leak strategic insights, or worse, be misinterpreted by those without adequate expertise. Others raise liability fears: the more you reveal, the more you expose yourself to legal claims.

International models provide contrasting approaches. In Germany, for instance, the ethics commission on automated driving recommended early on that AV decision-making protocols be transparent and reviewable. This gave rise to proposals for centralised oversight authorities tasked with certifying AV logic. In China, AV systems are subject to mandatory data-sharing agreements with state authorities, raising its own concerns about surveillance and geopolitical leverage. The United States, by contrast, has largely taken a laissez-faire approach, relying on voluntary safety self-assessments submitted to the Department of Transportation—documents that, in practice, vary wildly in quality and are nonbinding. *The result is a fragmented landscape where the norms of disclosure are shaped less by regulation than by public relations strategy.*

This opacity doesn't only affect crash investigation. It undermines democratic deliberation over values. Take, for example, how AVs prioritise safety in tight scenarios. Should the car favour the lives of its passengers or pedestrians? Should it prioritise the young over the old? Should it take into account legality (e.g., whether a pedestrian is jaywalking) when calculating whom to protect? These are not merely technical questions; they are moral choices coded into the vehicle's logic. When such choices are made in secret, we lose the opportunity to contest them, revise them, or ensure they align with our collective values. *An algorithm that encodes moral trade-offs without public scrutiny risks cementing one corporation's ethics as de facto law on public roads.*

Trade secrecy, in this context, becomes more than a business tactic—it is a means of shaping the moral and legal infrastructure of mobility without consent. The idea that AVs are "neutral" actors is an illusion; every design decision reflects human priorities, and those priorities deserve scrutiny. Moreover, the assumption that greater transparency must come at the cost of innovation is a false binary. Other safety-critical industries—pharmaceuticals, aviation, nuclear energy—have managed to balance proprietary research with rigorous public safety oversight. *The challenge is not technical impossibility but political will.*

Importantly, transparency is not a synonym for total openness. It can take graduated forms. Companies might retain core IP while disclosing policy-level logic (e.g., the system's prioritisation protocol), model behaviour under defined test cases, or submit to independent third-party audits whose findings are published in public-facing reports. Similarly, regulators might impose differential disclosure requirements: requiring deeper transparency in post-incident investigations than in normal operations, or granting limited access to certified experts rather than full public release. These are not radical ideas—they are industry standards in other domains.

Still, the AV sector has a long way to go before these norms are established. Currently, much of the public's knowledge about AV performance comes from curated demos, company press releases, or selective data drops. When accidents occur, firms often shift attention to vague statistical reassurances ("millions of miles driven") rather than engage substantively with the incident's moral and technical stakes. In doing so, they forfeit an opportunity to earn trust—not just in their products, but in the very future they claim to be shaping.

The stakes of this opacity are rising. As AVs move from limited trials to broader deployment, the volume of edge cases will increase, and so too will the burden of public accountability. Transparency is not just about post-mortem diagnostics—it is about building legitimacy. A vehicle that decides when and how to brake in a crowd must be answerable not only to its engineers but to the society it navigates. Without that, autonomy risks becoming an unaccountable authority. And in that world, the steering wheel may be gone—but the question of who holds control remains more urgent than ever.

Chapter 4
Cities on Autopilot

A self-driving car does more than replace a human driver—it subtly reorders the logic of urban life. Often presented as a technological leap forward for personal mobility, it quietly instigates a radical reshaping of cities themselves. Autonomous vehicles do not simply enter pre-existing streets; the streets are remade around them. This remaking manifests in both overt and subtle ways: curbless intersections in test zones, roundabouts widened to accommodate machine-calculated turning radii, and the installation of edge-computing nodes tucked behind lampposts. Crosswalks are recalibrated to be "readable" by algorithms rather than pedestrians, and speculative renderings of smart neighbourhoods imagine streets that learn from vehicles rather than citizens. The future city is being designed from sensor grid inward, not human experience outward.

The implications for public space, civic life, and social equity are immense. Traditional urban planning has always negotiated between walkers, cyclists, transit users, drivers, and those who linger or trade along streets. AV-centric planning, however, begins with the machine's perspective. Streets are optimised for algorithmic readability, and human unpredictability—jaywalking, idling, meandering—is quietly recast as "noise." In this framework, spontaneity is an obstacle, not an asset. The more efficient the city becomes for autonomous mobility, the more inhospitable it risks becoming to human life lived in its fullest texture.

We are witnessing a new infrastructural determinism. Just as the automobile of the twentieth century shaped zoning laws, freeway expansions, and suburban sprawl, the AV era exercises influence differently but no less profoundly. Control is now exerted through sensor placement, HD mapping boundaries, machine-learning feedback loops, and digital right-of-way protocols. Power is not wielded by bulldozers alone; it is encoded in software, regulated through access to urban data layers. Whereas prior debates centred on concrete and cost, AV infrastructure raises questions of governance: Who owns the sensor data? Whose safety is prioritised in routing algorithms? What behaviours are rendered invisible by lidar and cameras?

A particularly insidious phenomenon is the rise of "invisible infrastructures"—alterations so subtle the public barely notices, yet which dramatically reshape urban life. Curb space may be reallocated from parklets and loading zones to AV staging areas. Dynamic lanes may shift automatically to favour autonomous vehicles, sidelining slower or unquantified users. These interventions recalibrate streets toward machine efficiency and predictability rather than human conviviality. People are no longer the primary reference point; they become variables to manage.

Layered onto this is the spatial logic of autonomy. AVs require clear sightlines, unambiguous intersections, and rigid traffic patterns. Streets once rich with improvisation—pop-up markets, mid-block conversations, slow-moving pedestrians—risk being reduced to paths of least resistance for vehicular logic. What benefits machine perception often constrains human spontaneity and imagination.

These changes do not affect all communities equally. Early AV infrastructure tends to concentrate in wealthier areas with high-quality mapping data, robust connectivity, and pilot programmes, while lower-income districts may serve as experimental zones, exposing residents to risk and disruption. Increased AV traffic can drive up rents, incentivise surveillance, and reorient street activity toward commercialised, tech-centric functions. "Smart" mobility thus risks entrenching existing inequalities under the guise of progress.

The rhetoric surrounding autonomous transport emphasises efficiency, optimisation, and innovation, yet rarely interrogates for whom these benefits accrue. When bus lanes are converted to AV corridors, is public mobility enhanced or curtailed? When sidewalks are narrowed for robot deliveries, what becomes of communal street life? Such questions are less technical than philosophical: how do we value friction, cohabitation, and slowness in the city? A frictionless street may please algorithms without serving citizens.

Promises of inclusivity—extending mobility to the elderly, disabled, or transport-poor—are similarly fraught. AVs designed for narrow user profiles may exclude those with different abilities or socioeconomic constraints. App-based services, credit requirements, and subscription models can marginalise those without digital access. Furthermore, deployment often favours predictable, mapped environments, leaving complex or chaotic neighbourhoods underserved. The result is a two-tiered system: frictionless movement for some, stagnation for others.

Underlying all of this is the question of agency. Who determines the shape of a city designed for autonomous mobility? Local governments in consultation with citizens, or private consortia negotiating behind NDAs? Public participation risks being replaced by corporate design logic. "Smart" often signifies intelligence flowing from proprietary datasets to machine learning, not from lived human experience. Rather than AVs adapting to cities, cities are adapting to AVs—with consequences far beyond traffic patterns.

This chapter explores the subtle yet profound ways urban life is being redefined by autonomous vehicles. It traces how the physical, ethical, and political architecture of cities bends toward automation: readable by machines, predictable for fleets, and efficient for service providers. It interrogates who gains, who is displaced, and what forms of public life are preserved or suppressed. The city on autopilot may be efficient, but is it still a city for people?

Ultimately, the stakes of AVs are not confined to dashboards—they are embedded in street design, access politics, and the quiet recalibration of everyday life. Delegating mobility to machines is not merely a matter of losing control over transport; it is the beginning of a loss of civic imagination itself, the ability to shape and inhabit shared urban spaces on human terms.

Infrastructure for Invisibility

Urban infrastructure has long been a physical manifestation of societal priorities—sidewalks for pedestrians, lanes for vehicles, parks for communal gathering. The advent of autonomous vehicles

(AVs) introduces a new imperative: the design of cities increasingly must cater not to human sensibilities but to machine perception. This shift entails the rise of what might be termed *invisible infrastructure*—elements embedded within the urban fabric not to be seen or touched by people, but to be detected and interpreted by sensors, cameras, and algorithms. While ostensibly enhancing efficiency and safety, these changes risk relegating pedestrians and other human-scale urban experiences to the margins, making their presence an obstacle to be managed rather than a central consideration.

The fundamental problem arises from the differing modes of perception between humans and machines. Pedestrians navigate cities using a complex blend of visual cues, social signals, and contextual understanding. Machines, however, rely on quantifiable inputs: LiDAR reflections, radar signals, GPS coordinates, and high-definition maps. For AVs to operate reliably, the environment must be legible to these sensors. That legibility often demands stark, simplified spaces—clear lane markings, standardised signage, well-demarcated crossing points. *Irregularities, ambiguities, and spontaneous human behaviours become noise to be filtered out.*

This dynamic transforms urban space into a binary field: visible or invisible, legible or illegible to machine eyes. Pedestrian crossings, for instance, may be re-engineered not primarily to enhance human convenience or safety, but to maximise sensor detectability. Tactile paving, contrasting colours, and embedded RFID tags serve dual functions—facilitating access for people with disabilities, yes, but more crucially ensuring that the AV's software can reliably predict pedestrian behaviour. Such design shifts may marginally improve safety, yet they impose a new kind of surveillance and control, constraining pedestrian freedom within a system optimised for machine interpretation.

Beyond crossings, the very fabric of streets is being reimagined. Traditional curbside functions—loading zones, bus stops, informal markets—are streamlined or relocated to reduce unpredictability. Physical obstructions are removed; temporary encroachments discouraged. Even street furniture like benches or bike racks are reconsidered through a lens of sensor clarity. The city becomes a stage tailored to a robotic performer, with choreography designed for precision rather than spontaneity. In doing so, *the city's social texture risks erosion.*

The consequences of such "invisibility infrastructure" extend well beyond aesthetics. Pedestrians who do not conform to standardised behaviour—children running unpredictably, street vendors stepping into the road, elderly individuals with slower gait—may be inadvertently penalised by AV systems calibrated for normative movement patterns. This dynamic privileges certain users while rendering others peripheral or even hazardous in the machine's calculus. The pedestrian is not merely a user of urban space but increasingly a variable in an algorithmic risk assessment—a presence to be anticipated, controlled, or occasionally excluded.

Cities experimenting with AV corridors or pilot zones illustrate these tensions vividly. In some North American cities, road diets that narrow lanes to calm human drivers are reversed in AV test zones to create wider lanes facilitating smoother autonomous navigation. Pedestrian refuge islands may be shortened or removed to simplify crossing detection. These modifications improve vehicle flow but at the expense of human comfort and safety perception. Residents in these zones often report feeling alienated from spaces that once welcomed them, replaced by corridors engineered primarily for machine transit.

The push for invisibility also intersects with digital infrastructures—networks of sensors, cameras, and communication nodes that silently map and monitor urban life. These layers form an invisible mesh whose primary purpose is to serve AVs' decision-making needs. However, such surveillance infrastructure, while ostensibly benign, raises concerns about privacy, data security, and civic agency. *Pedestrians become data points tracked constantly, their movement patterns analysed not just for traffic optimisation but potentially for law enforcement or commercial profiling.* This fusion of physical and digital invisibility embeds a new regime of urban governance—less visible yet profoundly influential.

Critically, the prioritisation of AV-friendly infrastructure risks deepening pre-existing inequalities. Well-resourced districts may receive early investments in smart infrastructure, while marginalised neighbourhoods are left with outdated, inconsistent, or unsafe urban fabric. This uneven distribution replicates and exacerbates socio-spatial divides, effectively rendering some pedestrian populations "invisible" not only to AV sensors but to planners and policymakers. The invisibility of pedestrians is both literal and metaphorical, reflecting broader patterns of exclusion.

Yet there are attempts to mitigate these effects through design innovation. Concepts like "shared space" seek to blur the boundaries between pedestrian and vehicle, encouraging cautious cohabitation rather than segregation. Sensor technologies are advancing to better recognise diverse human behaviours, including gestures and group dynamics. Some cities are exploring participatory design processes to ensure that AV infrastructure respects and enhances local patterns of movement and interaction. These efforts highlight a vital insight: *infrastructure need not render humans invisible; it can be designed to make machines more human-aware.*

The challenge lies in reconciling competing demands. AV developers require environments optimised for reliability, predictability, and efficiency. Urbanists and community advocates demand spaces that nurture social interaction, safety, and inclusivity. The tension is not trivial—it is foundational to how cities will evolve in an automated future. Invisible infrastructure is not merely a technical condition; it is a political and ethical choice about whose presence matters, whose behaviours are normalised, and how urban life is orchestrated.

As autonomous vehicles increasingly navigate our streets, the infrastructure supporting them will shape more than traffic patterns; it will influence the rhythms of everyday life. *When urban design prioritises the needs of machines over people, it risks creating cities where human presence is tolerated rather than celebrated—where the vibrant unpredictability of street life is replaced by the cold precision of sensor-readability.* Recognising and challenging this shift is essential for crafting cities that remain vibrant, equitable, and truly alive in the age of automation.

The Hidden Geometry of Streets

Cities are living organisms shaped not only by their inhabitants but by the invisible geometries underpinning their form and function. Streets, often dismissed as mere conduits for movement, constitute the veins of urban life, orchestrating a complex interplay of spatial relationships, social encounters, and cultural rituals. The advent of autonomous vehicles (AVs) introduces a tectonic shift in this geometry—an algorithmic imperative that redefines the city's spatial order and social fabric in subtle yet profound ways. As streets are recalibrated to serve the logic of machines, the question emerges: what happens to the human rhythms that once thrived there?

At first glance, the reconfiguration of streets for AVs appears a technical necessity. These vehicles rely on highly precise sensor data, standardised signalling, and predictable environmental cues to navigate safely. Sharp turns, ambiguous intersections, and irregular road markings create challenges for machine vision systems; thus, urban designers are increasingly tasked with transforming streets into legible, streamlined environments. This "machine-readable" city demands a geometry of uniformity—wide lanes to accommodate sensor field-of-view, dedicated corridors free of obstacles, and intersection designs that minimise unexpected behaviour. The result is a city that privileges spatial clarity over complexity, precision over improvisation.

This transformation operates on multiple scales. On the micro level, the shape of curbs, the placement of bollards, and the texture of crosswalks are calibrated to enhance sensor detection rather than human comfort. Curbless intersections or simplified "shared spaces" may be designed to reduce physical clutter, but they also reduce the tactile cues pedestrians use to navigate and negotiate safety. At the meso level, the street grid may be reorganised to facilitate smooth flow of AV fleets—modifying traffic signal timings, introducing one-way routes, or redesigning roundabouts to conform to optimal turning radii. On the macro level, new zoning regimes may emerge, where AV corridors, geofenced zones, and service areas impose invisible boundaries on urban mobility.

Yet these physical reconfigurations carry symbolic weight. Streets have always been places of encounter, contestation, and expression—spaces where urban citizens interact, assert presence, and create meaning. By reshaping streets for AVs, cities risk stripping away these layers of sociality. The hidden geometry of autonomy tends to flatten the spatial hierarchy that nurtures urban vitality: informal gathering spots may be replaced by "smart" seating designed for efficiency; vendor stalls displaced by loading bays calibrated for robot deliveries; pedestrian wait times programmed to favour vehicle throughput over human patience. *The geometry of streets becomes a geometry of exclusion, optimising for flow at the expense of encounter.*

This shift also challenges the delicate balance between predictability and surprise that characterises vibrant public spaces. AVs demand predictability—uniform speeds, clear lines of sight, rule-abiding road users. However, human streets thrive on unpredictability: children darting between parked cars, street musicians pausing to engage passersby, neighbours pausing for spontaneous conversation. These moments, often dismissed as inefficiencies in traffic engineering, are vital to the social fabric. As streets are engineered to suppress such unpredictability for machine safety, *the city risks losing its soul—the serendipity that animates daily life.*

Real-world examples of these tensions abound. In cities like Phoenix and Pittsburgh, AV pilot zones have introduced simplified intersections and widened lanes. While these adjustments improve algorithmic navigation, residents report a decline in pedestrian friendliness. Crossing streets becomes less intuitive; the rhythmic flow of foot traffic is interrupted by prioritised vehicle paths. Similar patterns emerge in Singapore, where test areas for autonomous shuttles have restricted pedestrian access during operating hours, effectively segregating human and machine circulation. The cost is not only convenience but a subtle erosion of public space as a shared social commons.

Moreover, the hidden geometry of AVs intersects with socioeconomic divides. Affluent districts are more likely to receive early infrastructure upgrades—improved road surfaces, sensor installations, and digital mapping—tailored for autonomous mobility. Conversely, under-resourced neighbourhoods may face neglect or exclusion from these innovations, reinforcing patterns of spatial injustice. The physical reconfiguration of streets, thus, becomes entangled with questions of equity: who is entitled to walk freely, who is managed as a risk, and whose movements are streamlined or constrained? *The street's geometry is political geometry.*

Yet, not all implications are bleak. The demand for AV-friendly streets also offers an opportunity to rethink urban design through a new lens—one that integrates human-centred principles with technological innovation. Some urbanists advocate for "hybrid geometries" that balance the needs of machines and people: dynamic curb spaces that adjust to varying uses, sensor-augmented crosswalks that communicate directly with vehicles and pedestrians, and street layouts that preserve human-scale permeability within AV-optimised grids. Such designs seek to re-humanise streets by embedding technology in ways that amplify, rather than suppress, social interaction.

Another emerging idea is to leverage the data richness of AV systems for adaptive urbanism—where street configurations are not fixed but respond in real time to patterns of use, pedestrian density, and community events. Imagine streets that modulate vehicle flow during festivals, expand pedestrian zones during rush hours, or signal to AVs to reduce speeds near playgrounds. This responsive geometry, while technologically complex, promises a form of coexistence rather than displacement between human unpredictability and machine precision.

Philosophically, the challenge is to resist the temptation of seeing streets purely as conduits for movement or data streams. Streets are also *stages for civic life*, repositories of collective memory, and sites of identity formation. The hidden geometry imposed by AVs risks transforming them into sterile corridors, where human presence is an afterthought. To preserve the city's vibrancy, designers, policymakers, and technologists must acknowledge that *the geometry of streets is inseparable from the geometry of social relations.*

As AVs assert their presence on city streets, we must remain vigilant to how their spatial logic shapes more than traffic flow. The lines painted on asphalt, the shapes carved into curbs, and the signals embedded in infrastructure are all part of a silent reprogramming—one that configures not just vehicles but urban lives. To engage with this reality requires moving beyond the veneer of technological progress to confront the ethical and social dimensions of urban form. Only then can cities retain their humanity amidst the promise and peril of automation.

Smart Cities or Segregated Zones?

The promise of autonomous vehicles (AVs) is often woven into the broader narrative of "smart cities"—urban landscapes where technology optimises every facet of life, from traffic flow to energy use, creating seamless, efficient, and inclusive environments. Yet beneath this optimistic veneer lurks a critical question: will the deployment of AV-centric infrastructure and services deepen existing socioeconomic divides, or even create new ones? The reality of AV rollouts thus far suggests a complex and troubling dynamic, where the technologies intended to democratise mobility risk entrenching patterns of exclusion and spatial segregation. The line between a truly smart city and a fractured urban patchwork is razor-thin, shaped by how access, data, and design are governed.

At its core, the issue revolves around *who benefits* from AV technology and *who is left behind.* Early deployments and pilot programmes for AV fleets have overwhelmingly favoured affluent, well-mapped, and technologically equipped neighbourhoods—districts with reliable cellular networks, existing digital infrastructure, and predictable traffic patterns. These areas offer the ideal testing ground for complex autonomous systems, with fewer variables to confound machine learning. In contrast, many lower-income or marginalised communities—already grappling with inadequate transit options—face delayed or no access to AV services. The result is a bifurcated city: parts transformed into sleek, sensor-laden corridors optimised for autonomous flow, others relegated to the slow lane of infrastructural neglect.

This geographic and economic disparity is not accidental but emerges from the underlying incentives driving AV deployment. Private companies dominate the sector, operating within capitalist frameworks that reward market viability and scalability over equitable service. Wealthier districts offer higher return on investment through greater ridership, premium service subscriptions, and safer, more controlled environments that reduce operational risk. Meanwhile, regions characterised by infrastructural complexity, informal street economies, or unstable networks represent costly challenges. Without regulatory frameworks mandating universal coverage, companies naturally concentrate resources where profit and predictability converge. *Smart city ambitions collide with market logics, producing zones of privilege and zones of invisibility.*

Beyond mere access, the spatial configuration of AV services threatens to exacerbate urban segregation. Cities are already marked by socioeconomic zoning—residential enclaves divided by income, race, and opportunity. AV corridors and geofenced "mobility zones" risk reinforcing these divides by privileging certain areas for high-quality autonomous transit while others remain tethered to legacy systems or underfunded public transport. In some pilot projects, ride-hailing fleets have been explicitly confined to specific "smart" districts, creating literal mobility islands within the city. Residents outside these zones encounter reduced mobility options and increased isolation. Far from bridging divides, AV infrastructure can harden urban fractures.

Data governance compounds this segregation. AVs generate and rely on enormous volumes of geospatial and behavioural data to navigate, optimise routes, and refine algorithms. However, data ownership is predominantly corporate, with limited public oversight or mechanisms for community benefit. Consequently, wealthier districts that host extensive AV operations become data-rich environments, informing targeted improvements, commercial opportunities, and municipal planning. Conversely, data deserts emerge in underserved neighbourhoods, where limited AV presence equates to fewer digital signals captured and less investment attracted. *In this emerging data economy, spatial inequity is doubled: both physical and informational divides deepen.*

The dynamics of pricing and service models further amplify exclusion. Many AV services operate on subscription or on-demand pay-as-you-go models, inherently privileging users with disposable income, smartphone access, and digital literacy. Marginalised populations—those relying on cash transactions, undocumented residents, or people with disabilities—may find these systems inaccessible or prohibitively expensive. Attempts to extend service via subsidised rides or shared fleets often fall short, constrained by regulatory hurdles, inadequate funding, or the design of technology ill-suited for diverse needs. This leaves a troubling picture where the promise of inclusive mobility is undermined by systemic barriers. *Mobility becomes a commodity rather than a right.*

One illustrative case is the deployment of AV shuttles in suburban testbeds. While touted as solutions to "last-mile" connectivity, these projects often skirt deeper transit inequities. Suburbs typically host residents with private vehicles, and such pilot programmes rarely serve densely populated, transit-dependent urban cores where mobility challenges are greatest. As a result, AVs risk enhancing convenience for those already mobile while leaving behind those who rely on buses, walking, or cycling. This dynamic replicates a classic pattern of technological "leapfrogging," where innovations bypass those most in need.

The governance of AV integration thus emerges as a critical determinant of whether these technologies widen or narrow divides. Cities with proactive, equity-focused policies—mandating minimum service coverage, integrating AVs with public transit, regulating data sharing, and subsidising access—stand a better chance of harnessing AVs for inclusive urban mobility. Yet, such frameworks are the exception rather than the rule. Many municipalities remain ill-equipped, under-resourced, or politically reluctant to impose equity mandates on a sector eager for rapid innovation and private control.

Equally important is the need for participatory design in AV deployment. Decisions about where and how AVs operate should not be dictated solely by corporations or technocrats but include meaningful input from affected communities. Residents of marginalised neighbourhoods often have intimate knowledge of local mobility needs and barriers that cannot be captured by data alone. Their voices are essential in shaping services that are not only technologically sound but socially just. Without such engagement,

AV rollouts risk replicating patterns of exclusion that have long characterised urban infrastructure.

Philosophically, the debate around smart cities and segregation hinges on how we conceive of urban justice. Is a smart city one that maximises efficiency and innovation regardless of who benefits, or one that deliberately prioritises equitable access and redistribution? The allure of technology-driven solutions must not obscure the political choices embedded in design and deployment. AVs, as embodiments of both technological possibility and market forces, crystallise this tension. *A city governed by algorithms without attention to social context risks becoming a patchwork of privileged enclaves and disenfranchised zones—a fragmented rather than integrated whole.*

Yet, recognising these risks also opens pathways for reform. Policymakers can craft regulatory frameworks that incentivise equitable service distribution, require transparency in data practices, and embed social impact assessments into AV project approvals. Public-private partnerships can align profit motives with community goals. Investment in digital infrastructure can be targeted to reduce disparities. And crucially, urban planners and technologists can prioritise multimodal, human-centred mobility ecosystems rather than singular AV-centric visions.

Ultimately, the question posed by AV-driven urban transformation is not only about technology but about the kind of cities we want to live in. *Will smart cities be inclusive spaces that empower all residents, or will they become segregated zones marked by unequal access, surveillance, and exclusion?* The answer depends not on inevitability, but on deliberate choices—about design, governance, and values—that will shape the future of urban mobility for generations.

Who Gets a Ride?

The promise of autonomous vehicles (AVs) is often framed in broad strokes of progress and liberation—ushering in an era where mobility is universally accessible, accidents plummet, and traffic inefficiencies vanish. Yet beneath this optimistic vision lies a critical and underexamined question: who, in fact, benefits from AV rollouts, and who risks being left behind? As with many technological revolutions, the promise of transformation often masks a landscape shaped by uneven access, social stratification, and infrastructural bias. The question of who truly "gets a ride" in the autonomous future exposes fault lines in urban planning, economic models, and social equity.

The earliest AV deployments have typically favoured environments and user groups that fit within narrow technical and commercial parameters. Wealthy, tech-savvy urban districts with robust digital infrastructure—think Palo Alto, Singapore's financial district, or certain European city centres—have become the primary beneficiaries. These areas offer predictable road conditions, consistent network connectivity, and a consumer base inclined towards early adoption of high-tech services. The result is a kind of geographic privilege, where residents enjoy cutting-edge mobility services, reduced commute times, and safer streets. In contrast, sprawling suburbs, informal settlements, and economically disadvantaged neighbourhoods are often excluded from initial AV zones due to technical challenges or lower profitability.

This spatial disparity is compounded by economic barriers. Many AV services are initially offered through premium-priced ride-hailing or subscription models. While these services promise convenience and efficiency, their cost structure inherently favours higher-income users who can afford frequent rides or monthly fees. Lower-income individuals, including essential workers, students, and the elderly on fixed incomes, may find AV options financially out of reach. Without targeted subsidies or inclusive pricing policies, autonomous mobility risks becoming yet another luxury good rather than a universal public utility. *The democratization of mobility depends not only on technology but on deliberate economic design.*

Beyond geography and income, the question of who gets a ride touches on physical and cognitive accessibility. AVs have the potential to transform mobility for people with disabilities, older adults, and others whose transportation options are limited. Yet, to fulfil this promise, vehicles and services must be designed with inclusive features—ramp access, adaptable interfaces, support for service animals, and accommodations for sensory or cognitive impairments. Early AV models, however, have largely focused on mainstream users, reflecting market-driven priorities rather than universal design principles. The risk is that AV rollouts may replicate existing accessibility gaps unless regulators and manufacturers prioritise inclusive standards from the outset.

Furthermore, the digital architecture underpinning AV services imposes additional hurdles. Access typically requires smartphones, stable internet connections, and digital literacy. Populations lacking these—whether due to economic, educational, or generational factors—may find themselves unable to navigate or afford AV platforms. This digital divide compounds physical mobility inequities, leaving vulnerable groups further marginalised. For residents in digitally underserved areas, or those unfamiliar with app-based services, the autonomous promise remains abstract, even elusive.

The interplay between public and private provision further complicates the picture. Many AV deployments are spearheaded by private companies seeking to capture market share rather than fulfil public service obligations. This commercial orientation shapes service availability, prioritising high-demand routes, affluent clientele, and profitable time slots. Public transit systems—already strained and underfunded in many cities—may be displaced or undercut by private AV operators, who siphon off riders without necessarily serving the broader community. Without strong regulatory frameworks, this dynamic risks exacerbating transportation deserts, where the most vulnerable populations see reduced options and increased isolation.

Nonetheless, some pilot projects and policy initiatives illustrate pathways towards more equitable AV access. For example, Helsinki's public transit authority has integrated autonomous shuttles into its multimodal network, prioritising routes that serve peripheral and underserved neighbourhoods. Similarly, the city of Las Vegas has partnered with AV providers to offer subsidised rides to elderly and disabled residents. These examples highlight the potential for hybrid models where public interest shapes service deployment, ensuring mobility enhancements reach beyond the economically privileged. Such approaches underscore the importance of policy in bridging technological promise with social reality.

The question of who gets a ride is not merely a logistical or economic challenge—it is a fundamentally ethical one. Mobility is deeply tied to autonomy, opportunity, and social inclusion. When access to transportation is uneven, so too are access to employment, healthcare, education, and civic participation. Autonomous vehicles have the potential to reshape these dimensions dramatically, but only if the rollout consciously prioritises equity. The risk is a fractured mobility landscape where technology amplifies existing disparities rather than redressing them.

Moreover, the implications extend beyond individual users to the collective urban fabric. If AV services cluster in affluent areas, they may contribute to increased spatial segregation and gentrification, driving up property values and displacing long-term residents. Conversely, neglected areas may suffer from reduced investment, diminished economic vitality, and entrenched poverty. Mobility is not an isolated good; it is interwoven with housing, employment, and social networks. *Who gets a ride is thus a proxy for who belongs, who participates, and who thrives in the city.*

As AV technology matures and scales, its governance will be crucial in shaping inclusive outcomes. Cities and nations must consider regulatory measures that mandate service coverage minimums, protect affordability, enforce accessibility standards, and promote data transparency to identify and correct disparities. Public engagement is vital to understand community needs and build trust in new mobility models. Without these interventions, the autonomous future risks becoming a technocratic landscape defined by privilege rather than a truly shared urban commons.

In navigating these complex terrain, one principle stands clear: *mobility equity is not an accidental byproduct of innovation; it is a deliberate design and policy choice.* Autonomous vehicles offer a transformative opportunity to reimagine urban movement, but realising this potential demands an unflinching focus on who benefits, who is excluded, and how technology can serve—not supplant—the social contracts that underpin just cities.

Chapter 5
The Labor Fallout

The arrival of autonomous vehicles signals a disruption far deeper than the mechanics of movement; it reaches into the very fabric of labour. Driving, long a cornerstone of global employment and personal identity, now confronts an uncertain horizon. The familiar image of a driver—behind a cab, a delivery van, or a long-haul truck—faces rapid obsolescence. This chapter interrogates the human consequences hidden behind the technical marvels of automation. Roads and cities are not the only systems being rewired; entire communities, economies, and individual senses of purpose are being reshaped, often abruptly and unevenly.

At first glance, the most visible impact is job loss. Taxi drivers, truckers, couriers, and delivery workers—professions that have provided stable employment for generations—are under direct threat. Unlike previous technological shifts, which tended to displace tasks incrementally, autonomous vehicle technology promises wholesale removal of the driver role. The scale is immense: in the United States alone, more than 3.5 million professional drivers could face redundancy. Across the globe, the numbers expand across multiple sectors and transport modalities. Yet the ripple effects go further still. Entire communities built around driving—support networks, small businesses, family economies—face cascading disruption. Driving is more than work; it is identity, culture, and social cohesion in countless regions.

Still, the picture is not one of unmitigated disappearance. Some roles will evolve rather than vanish entirely, producing a complex landscape of displacement and transformation. New opportunities will arise in remote vehicle supervision, fleet management, data analysis, vehicle maintenance, and human-machine interface design. Yet these roles demand skills, education, and adaptability often absent among those most directly displaced. Access is far from guaranteed. The labour market shift is not merely quantitative; it is profoundly qualitative, reshaping livelihoods, social standing, and prospects for intergenerational mobility.

Beyond economics lies the psychological dimension. Driving is more than a sequence of actions; it is a source of autonomy, pride, and professional competence. The sudden erosion of these roles can provoke identity crises, loss of purpose, and diminished dignity. For many, driving is a craft, a daily rhythm intertwined with social engagement. To be replaced by a machine is to confront obsolescence in the very skills that define one's self-worth, with consequences that reach into mental health and social marginalisation. Automation does not merely alter labour—it reshapes the human psyche.

These pressures intersect with labour organisations and policy debates. Unions and worker groups struggle to defend the interests of displaced or transitioning workers, while the globalised nature of AV technology complicates collective bargaining. Gig economy models, fragmented labour markets, and cross-border deployment introduce new barriers to representation. Policymakers often lag behind technological disruption, constrained by regulatory inertia, competing economic priorities, and insufficient safety nets. The result is a widening gap between the pace of innovation and the protection of workers, leaving many exposed not just to unemployment, but to the erosion of meaningful, dignified work.

Understanding the labour implications of AVs demands a holistic lens—one that blends economic data, sociological insight, and ethical reflection. We must ask not only which jobs will exist tomorrow, but what forms of life those jobs will sustain. How can education, policy, and corporate responsibility converge to provide pathways rather than barriers? How do we prevent automation from deepening inequality, fracturing communities, and undermining dignity? These are the questions this chapter seeks to illuminate, presenting a nuanced view of labour at the intersection of innovation and justice.

By confronting the full complexity of AV-induced labour change, we move beyond simplistic narratives of loss or progress. The human stories behind algorithmic efficiency, the structural forces shaping workforce transformation, and the moral responsibilities they entail are central to understanding the broader stakes of the autonomous revolution. Labour disruption is not a sidebar to technological advancement; it is a defining feature, demanding careful thought if we are to shape a future in which technology serves all, rather than a privileged few.

The End of Driving as Work

The sight of a driver behind the wheel—whether ferrying passengers, hauling goods, or delivering essentials—has for generations been a ubiquitous emblem of the modern economy. Driving is not merely a means of transportation but a profession that sustains millions worldwide, especially within working-class communities and regions where alternative employment opportunities are scarce. As autonomous vehicle (AV) technologies move from experiment to everyday reality, the traditional role of the driver faces an existential threat. This seismic shift in labour markets extends far beyond individual job loss, triggering profound ripple effects across families, local economies, and social structures that have long depended on driving as a source of income and identity.

The scale of potential job displacement is unprecedented. In the United States, the Bureau of Labor Statistics estimates that approximately 3.5 million people work as heavy and tractor-trailer truck drivers, delivery drivers, taxi drivers, and chauffeurs. Globally, these numbers multiply exponentially. Trucking alone accounts for a large share of goods movement, with millions employed in logistics, long-haul transport, and last-mile delivery. Autonomous technology promises to eliminate many of these roles by removing the need for a human operator—especially on highways and repetitive routes where automation excels. Unlike incremental changes seen in earlier mechanisation waves, AVs threaten to render entire occupations obsolete within a compressed timeframe.

However, the implications are not limited to the drivers themselves. Driving jobs often form the backbone of local economies. In many regions—rural towns reliant on trucking routes or urban neighbourhoods with clusters of taxi drivers—the loss of driving employment undermines economic vitality. Small businesses servicing these workers, from vehicle maintenance shops to diners and retail outlets, may also experience decline. Families dependent on steady driving incomes face sudden insecurity, intensifying social stresses. The effects compound in communities where driving is intertwined with social identity, cultural belonging, and intergenerational aspirations. *Losing driving jobs thus destabilises economic ecosystems and frays social cohesion.*

The technological transition is uneven and unevenly felt. While autonomous freight may phase in gradually along highways first, urban passenger transport confronts unique challenges. Yet ride-hailing drivers and delivery couriers in cities are already facing reduced demand or altered job conditions due to AV trials and pilot programmes. The gig economy's precarious employment structures amplify vulnerability, with many workers lacking formal protections or access to retraining. Informal or undocumented drivers may be particularly exposed, lacking social safety nets. *The end of driving as work risks deepening existing labour market inequalities rather than mitigating them.*

Not all driving jobs are identical, and the differential impacts reflect this diversity. Long-haul trucking, characterised by extended solo drives and relatively predictable routes, is seen as particularly vulnerable to automation. Conversely, roles demanding complex urban navigation, customer interaction, or rapid decision-making are initially less susceptible, though the technological frontier is advancing rapidly. Moreover, some segments of the workforce may transition into new roles related to AV operation, maintenance, or supervision. Yet this transition demands skills and education that many displaced drivers do not possess or cannot readily acquire. *The labour market's capacity to absorb displaced workers into new, meaningful employment remains uncertain and uneven*

The loss of driving jobs also implicates questions of identity and dignity. Driving is often more than a paycheck—it is a craft, a source of autonomy, and a marker of social status within many communities. The shift from human to machine control challenges these personal and collective narratives. Workers report feelings of obsolescence, frustration, and loss as their expertise is rendered redundant by automation. These psychological dimensions have ripple effects on mental health, community morale, and social stability. *The end of driving as work is, therefore, not just an economic phenomenon but a cultural and existential rupture.*

Policy responses to this impending labour fallout have been slow and fragmented. Governments and labour organisations struggle to anticipate and mitigate the scale of disruption. While some pilot retraining programmes and social support measures exist, they are often limited in scope and funding. The pace of AV development outstrips regulatory capacity, creating a governance gap. Unions representing drivers face complex challenges: declining membership, fragmented gig work structures, and the need to negotiate with both traditional employers and emerging tech firms. These dynamics complicate collective action and the articulation of workers' rights in a rapidly evolving labour landscape.

Internationally, responses vary widely, reflecting differing economic structures, labour protections, and political priorities. In countries with strong social welfare systems, safety nets may soften the transition, while in others the fallout could be harsher and more immediate. The globalised nature of supply chains and AV technology development further complicates the picture, as job displacement in one region may be offset by emerging opportunities elsewhere—or by intensified competition for scarce positions. *The labour fallout from AVs is thus a transnational challenge demanding coordinated, cross-sectoral policy innovation.*

Amid these challenges, some voices advocate for a reimagining of work and social value beyond traditional employment. Concepts such as universal basic income, job guarantees, and lifelong learning have gained traction as potential buffers against automation-induced displacement. Others highlight the need for anticipatory governance frameworks that integrate technology assessment with social policy, ensuring that workers are not left at the mercy of market forces alone. These debates underscore the broader societal imperative: how to navigate the transition from human-centred labour to a hybrid human-machine economy without forsaking social justice.

The trajectory of driving as work illuminates a broader truth about automation's disruptive potential: technological progress does not guarantee social progress. *Without deliberate intervention, the benefits of AVs risk accruing to capital owners and technology elites while eroding the livelihoods and dignity of millions of workers.* Understanding the multifaceted impacts on driving sectors—and by extension, dependent communities—is essential to crafting responses that are both economically sound and ethically grounded.

As the world accelerates toward autonomous mobility, the end of driving as work stands as a stark reminder that innovation reshapes not only cities and roads but the very foundations of labour, identity, and community. This chapter invites readers to confront these realities with rigor and empathy, exploring the intricate interplay of technology, economy, and human experience at a pivotal moment in labour history.

Job Loss vs. Job Shift

The rise of autonomous vehicles (AVs) presents one of the most transformative disruptions to the labour market in recent history, particularly for occupations rooted in driving. Headlines frequently focus on the spectre of widespread job loss, conjuring images of millions displaced from steering wheels. Yet the reality is more nuanced. While automation will undoubtedly eliminate many driving roles, it will simultaneously create, reshape, or redefine others. Understanding this distinction—between job loss and job shift—is essential for grasping the full impact of AVs on work, livelihoods, and economic structures.

Job loss, in the context of AVs, refers to the outright disappearance of roles rendered redundant by technology. Heavy goods vehicle drivers, taxi drivers, and delivery personnel stand at the forefront of this risk, especially in tasks involving repetitive, long-distance, or predictable routes. For example, autonomous trucks operating on highways with minimal human intervention are expected to displace many long-haul drivers. Similarly, robotaxis promise to supplant human drivers in urban ridesharing services, reducing reliance on human labour. The scale of this displacement is significant: estimates suggest that millions of driving jobs worldwide could vanish as automation matures. Such losses bear profound economic and social consequences, particularly for communities where driving employment is a mainstay.

However, framing the AV labour impact solely as job loss overlooks the parallel phenomenon of job shift—where roles evolve in nature rather than vanish outright. Automation frequently transforms tasks rather than replacing entire occupations wholesale. In the AV ecosystem, this means that while direct driving functions diminish, new roles emerge that centre on managing, monitoring, and maintaining autonomous fleets. These include remote vehicle supervisors who oversee multiple AVs, technicians specialising in sensor calibration and hardware maintenance, data analysts interpreting vast streams of vehicle-generated information, and customer support professionals handling interactions in AV service contexts.

Such transformations suggest a partial reconfiguration of the workforce. For instance, the truck driving industry could see a rise in logistics coordinators responsible for integrating AVs into supply chains, or "mobility managers" who orchestrate dynamic fleet deployments based on real-time demand. These roles often require higher levels of technical proficiency, problem-solving skills, and familiarity with digital platforms. The transition from manual driving to supervisory and technical work thus shifts not only the nature of labour but also its skill demands.

Yet this shift presents challenges of access and equity. The workers most vulnerable to displacement—often lower-skilled, less formally educated, or economically marginalised individuals—may lack the resources or opportunities to acquire new competencies demanded by emerging roles. Without robust retraining programmes, educational pathways, and supportive policies, the job shift may result in structural unemployment for significant segments of the driving workforce. *The qualitative change in labour thus risks deepening existing socioeconomic inequalities unless accompanied by deliberate intervention.*

Furthermore, the distinction between job loss and job shift is not always clean or immediate. Many workers will experience transitional uncertainty, juggling diminished driving opportunities with precarious or temporary roles in the new AV economy. Hybrid job models may arise, such as semi-autonomous driving positions requiring intermittent human intervention, creating ambiguous labour categories with uncertain rights and protections. The gig economy, already pervasive in ridesharing and delivery sectors, further complicates these dynamics, blending full-time, part-time, and contract work in unstable configurations. This blurring of employment status exacerbates vulnerabilities and challenges traditional labour frameworks.

Global case studies illuminate these complexities. In the United States, the trucking industry is already witnessing trials of remote driving support, where drivers manage multiple autonomous trucks from control centres. While such roles extend employment opportunities, they demand advanced technical training and may concentrate jobs in urban hubs, displacing rural drivers. In China, pilot programmes integrating AV taxis in metropolitan areas have spawned new categories of "fleet managers" and "user experience specialists," but questions remain about the inclusivity and scale of these roles. These examples highlight that job shifts depend heavily on local economic structures, labour markets, and policy environments.

Beyond economic and logistical considerations, the qualitative nature of work also evolves. Jobs transformed by AVs may alter workers' sense of purpose and identity. For many, driving is not merely a task but a source of autonomy, pride, and social engagement. Transitioning to supervisory or technical roles changes daily experiences, workplace cultures, and even status hierarchies. The psychological adjustment required can be profound, necessitating attention to worker well-being and support systems alongside retraining efforts.

The policy implications of job loss versus job shift are equally significant. Mitigating the social costs of displacement demands more than passive acceptance; it requires proactive governance to facilitate skill development, enforce fair labour standards, and ensure inclusive access to emerging job categories. Strategies such as public investment in vocational training, partnerships between technology firms and educational institutions, and adaptive social safety nets are critical to enabling workers' transitions. Moreover, recognising the differentiated pace and geography of automation adoption is vital for tailoring responses that address local vulnerabilities.

The debate also challenges prevailing narratives about technology and progress. Automation is often portrayed as an unstoppable tide sweeping away obsolete jobs. Yet history teaches that labour markets adjust dynamically, creating new opportunities even as old ones vanish. The AV revolution follows this pattern, but with unprecedented speed and scale. The central question is whether societies will shape this transition to foster broadly shared prosperity or allow it to exacerbate exclusion and precarity.

In this context, the distinction between job loss and job shift transcends semantics. It frames a critical conversation about justice, agency, and the future of work in an automated world. The promise of AVs lies not only in technological sophistication but in how the accompanying labour transformations are managed, who is included, and how dignity and opportunity are preserved. As autonomous vehicles redraw the boundaries of work, understanding these shifts equips policymakers, industry leaders, and communities to navigate disruption with foresight and fairness.

By exploring the contours of eliminated and transformed roles, this chapter invites readers to confront the complex realities beneath headlines of job apocalypse. It underscores that the labour landscape of the AV era is not a monolith but a mosaic—comprised of losses, gains, adaptations, and challenges requiring nuanced, comprehensive responses. In doing so, it contributes to a deeper, more informed dialogue about technology's role in shaping equitable and resilient labour futures.

Automation's Psychological Toll

The advance of automation, especially within the realm of autonomous vehicles, marks a profound shift not only in the mechanics of work but also in the very nature of human identity tied to labour. The psychological consequences of this transition are as significant as the economic and structural changes it precipitates. For millions of professional drivers, labour is more than a transactional exchange—it is a source of self-worth, autonomy, social connection, and dignity. The gradual removal of the human driver from the vehicle's command seat challenges these deeply ingrained aspects of identity and raises urgent questions about how societies value work beyond productivity.

Driving has historically provided many workers with a palpable sense of agency and craftsmanship. Long-haul truckers, taxi drivers, and delivery personnel often describe their profession as a skilled trade, honed through experience and intuition. The capacity to navigate complex routes, respond to unpredictable situations, and maintain a connection with the road instils a form of pride and competence. With the advent of autonomous vehicles, these roles risk being reduced to obsolescence or relegated to remote supervision, eroding the hands-on engagement that underpins professional identity. This transition can provoke feelings of displacement, helplessness, and loss.

Research in occupational psychology underscores that work is a primary avenue through which individuals derive meaning and structure in their lives. The erosion of driving jobs threatens this foundation, potentially undermining mental health and well-being. Studies of workers displaced by automation in other industries—manufacturing, mining, and clerical work—have documented increased rates of anxiety, depression, and social withdrawal. In transportation sectors, where work is often solitary but meaningful, the psychological toll may manifest through diminished self-esteem and alienation. The loss extends beyond income; it is a loss of role, purpose, and identity within community and society.

Moreover, the shift from active driving to more passive or supervisory roles alters the experiential quality of work. Remote operators overseeing autonomous fleets confront different psychological demands—monitoring screens rather than roads, reacting to algorithmic alerts rather than direct sensory cues. This shift can induce what occupational psychologists term "cognitive disengagement" or "de-skilling," where workers feel disconnected from their tasks and less motivated. The change in work rhythms and social interactions can exacerbate feelings of monotony and invisibility. *Automation, in this sense, transforms not only what people do but how they feel about what they do.*

The psychological impact also intersects with broader social identities. In many cultures and communities, driving is entwined with notions of masculinity, independence, and social status. The disappearance of driving jobs can therefore destabilise collective identities and challenge traditional social roles. For example, truck driving in the United States and parts of Europe has been culturally celebrated as emblematic of freedom and rugged individualism. The removal of this role threatens more than employment—it threatens narratives of belonging and self-conception. This cultural dimension complicates the adaptation to automation, as displaced workers navigate not only economic uncertainty but existential dislocation.

The mental health effects of automation are further compounded by the precarity and unpredictability it introduces. As traditional driving roles diminish, displaced workers often face fragmented employment, reduced job security, and income volatility. These conditions are fertile ground for stress, anxiety, and social marginalisation. Without comprehensive support systems—including accessible mental health services, retraining programmes, and community engagement initiatives—the human costs of automation risk being overlooked or minimised. *The psychological toll is thus not an incidental byproduct but a central facet of the labour transition.*

Employers and policymakers bear a critical responsibility in recognising and mitigating these impacts. Interventions must go beyond economic compensation and retraining to include psychological support, identity reconstruction, and social reintegration. Programmes that facilitate peer networks, promote worker agency in transition processes, and acknowledge the emotional dimensions of displacement can foster resilience. In some pilot projects, companies have begun integrating mental health resources and participatory redesign of workflows to ease the shift. Yet such initiatives remain sporadic and insufficiently scaled.

Philosophically, the psychological consequences of automation call for a reevaluation of work's role in human flourishing. If technology renders certain forms of labour obsolete, societies must grapple with how to sustain dignity,, purpose, and social inclusion outside traditional employment paradigms. This invites debates about universal basic income, community-based work, lifelong learning, and redefining societal contributions beyond economic productivity. *The psychological toll of automation is a symptom of deeper questions about meaning and value in a post-industrial world.*

In the context of autonomous vehicles, these challenges acquire particular urgency. The driver's seat is not just a workplace but a stage for human skill, decision-making, and interaction. The transition to automation risks severing this link, with profound repercussions for individual and collective well-being. Understanding these psychological dimensions enriches the discourse on AV labour impacts, urging a holistic approach that respects both economic realities and human dignity.

As autonomous technology advances, the labour landscape it shapes will not only determine economic outcomes but also define the social and psychological fabric of work. Recognising and addressing the psychological toll of automation is essential for fostering a future where technological progress and human well-being advance in tandem—not at the expense of one another.

Union Battles and Policy Gaps

The rapid advance of autonomous vehicle technology confronts labour markets with an urgent paradox: as millions of driving jobs face displacement, organised labour and public policy often lag behind, struggling to mount effective resistance or offer meaningful protection. The emerging battles between labour unions and corporate interests are not just about wages or working conditions; they represent a fundamental contest over the future of work, power, and justice in an automated economy. At the same time, policy frameworks designed to safeguard displaced workers reveal significant gaps, leaving many vulnerable at a moment of unprecedented upheaval. Understanding these dynamics is critical to grasping the full societal implications of the AV revolution.

Historically, unions have served as vital advocates for workers navigating industrial transformations. In sectors such as trucking and taxi services, unions have long fought for fair pay, job security, and safe working conditions. However, the arrival of AVs disrupts established labour relations in profound ways. Automated vehicles threaten to render whole categories of drivers redundant, eroding union membership bases and diluting collective bargaining power. This dynamic is compounded by the increasing prevalence of gig economy models in ridesharing and delivery, where workers operate as independent contractors with minimal protections, circumventing traditional union structures. *Labour organisations thus face the double challenge of resisting displacement while reinventing themselves in a fragmented, digitalised labour market.*

One prominent example is the decline of truck driver unions in the United States and Europe. As companies invest in autonomous freight fleets, the bargaining leverage of drivers diminishes. Some unions have responded by demanding retraining guarantees, severance packages, or inclusion in transition planning. Yet these efforts often confront resistance from corporations focused on cost reduction and technological acceleration. Moreover, the international nature of AV development allows companies to relocate jobs or outsource supervisory roles to lower-wage regions, undermining local labour protections. *The transnational character of automation complicates union strategies, requiring new forms of coordination and solidarity.*

In urban passenger transport, the erosion of union power is even more acute. Traditional taxi drivers, who have historically benefitted from union representation, have seen their numbers dwindle as ridesharing platforms and autonomous pilot programmes proliferate. Gig workers in these sectors face precarious employment, with little access to benefits, collective bargaining, or job security. Attempts to unionise gig workers encounter legal and political obstacles, as firms classify them as independent contractors. This structural shift fragments the workforce and weakens labour's capacity to negotiate protections in the face of automation-driven displacement. *The battle for representation in the AV economy is as much about redefining labour categories as it is about technology.*

Policy responses to the looming labour fallout remain patchy and reactive. While some jurisdictions have introduced retraining initiatives, income support programmes, or transitional employment schemes, these measures are often limited in scale and scope. The pace of AV technological adoption outstrips regulatory frameworks, creating a policy lag that leaves workers exposed. Moreover, debates over liability, licensing, and data governance often overshadow labour concerns, reflecting a broader prioritisation of innovation and market growth over social protection. *This policy inertia risks replicating past patterns where technological change outpaces social adaptation, intensifying inequality and exclusion.*

Emerging labour movements are increasingly vocal in challenging this status quo. New coalitions of workers, activists, and scholars advocate for comprehensive frameworks that integrate automation policy with labour rights. Proposals include stronger regulation of gig work, expanded social safety nets, universal basic income pilots, and legally mandated retraining guarantees tied to automation deployment. These initiatives seek to shift the debate from reactive mitigation to proactive shaping of the future of work, emphasising dignity, inclusion, and shared prosperity. *The fight over AV labour futures is therefore also a struggle over political imagination and the scope of social justice.*

Yet these movements face formidable obstacles. The political influence of technology firms and logistics corporations, equipped with significant capital and lobbying power, often dwarfs that of fragmented labour constituencies. Regulatory capture, ideological commitments to market freedom, and global competition constrain government action. Additionally, the rapid pace of technological change complicates policymaking, as legislators grapple with uncertainty over timelines, impacts, and appropriate interventions. *The tension between innovation and regulation plays out not only in economic terms but in democratic governance and accountability.*

Internationally, the landscape varies considerably. Countries with robust labour protections and social welfare systems—such as those in Scandinavia—tend to have more comprehensive policy frameworks addressing automation's labour impacts. In contrast, jurisdictions with weaker unions and social safety nets, including many developing economies, face heightened risks of worker marginalisation and social unrest. Cross-border coordination on labour standards, technology governance, and worker protections remains nascent but increasingly necessary as AV supply chains and service platforms transcend national boundaries. *Globalisation thus adds layers of complexity to domestic union struggles and policy debates.*

The unfolding contest between unions, corporations, and policymakers around AV labour issues is emblematic of a broader societal reckoning with automation. It foregrounds questions about who holds power in shaping technological futures, whose interests are prioritised, and how societies negotiate the social consequences of innovation. Beyond economic metrics, it is a struggle over voice, agency, and democratic participation in the governance of work. As automated vehicles become embedded in everyday life, these labour battles will shape not only employment patterns but the contours of social cohesion and justice.

Recognising the urgency and stakes of these challenges compels a holistic approach that bridges technological foresight with labour advocacy and policy innovation. Supporting displaced workers demands integrated strategies encompassing retraining, social protection, collective representation, and democratic accountability. Furthermore, fostering new forms of worker solidarity adapted to the digital, automated economy is vital to counterbalance corporate power and ensure equitable outcomes. *The future of work in an autonomous world will be defined as much by political will and social organisation as by technological capability.*

In navigating the turbulent intersection of automation and labour, this chapter highlights how union battles and policy gaps are not peripheral but central to understanding the broader human and societal costs of AV adoption. It calls on stakeholders—governments, industry, labour, and civil society—to confront these issues with seriousness, creativity, and commitment. The outcome of these struggles will determine whether automation advances as a force for shared progress or deepened division.

Chapter 6
Surveillance in Motion

As autonomous vehicles (AVs) become an increasingly familiar presence on our streets, a subtler transformation unfolds alongside their quiet navigation: our movements themselves are converted into streams of data, continuously gathered, analysed, and, ultimately, commodified. This chapter explores the intricate surveillance embedded within the AV ecosystem, revealing that these machines are far more than transport devices—they are sophisticated sensing platforms, constantly translating motion into information. The shift from human-driven to machine-operated transport inaugurates a scale of monitoring unprecedented in peacetime, raising pressing questions about consent, privacy, and the commercialisation of mobility.

At the core of every AV is an insatiable demand for data. To negotiate roads safely and efficiently, autonomous systems maintain constant awareness of their surroundings, other road users, infrastructure, and environmental conditions. High-definition cameras, lidar, radar, GPS, and interconnected sensor networks collectively offer a panoramic perception that far exceeds human capabilities. Yet this observation extends beyond operational necessity: every trip generates detailed records of routes, speed patterns, stops, passenger behaviour, and even biometric indicators. Mobility, once ephemeral, is thus transformed into a persistent archive of personal and collective information, challenging conventional understandings of privacy and the very notion of informed consent.

The technological imperatives of AVs have given rise to a 'mapping arms race'. Control over precise, real-time geospatial data has become a coveted asset, both commercially and geopolitically. Corporations and governments alike seek to develop, maintain, and monopolise high-resolution digital maps that underpin AV functionality. These maps combine static infrastructure data with constantly updated dynamic inputs—traffic flows, construction zones, pedestrian movement—producing living cartographies of the urban environment. The stakes are significant: ownership, access, and governance of these maps influence not only transport efficiency but also wider questions of surveillance, civic control, and national security.

Equally consequential is the commodification of behavioural data derived from AV operation. Insurance firms, law enforcement agencies, and mobility service providers increasingly employ predictive analytics to assess risk, tailor services, and inform decision-making. While such approaches promise efficiency and personalisation, they also risk codifying bias and amplifying inequality. Those whose travel patterns deviate from normative models may face higher premiums or closer scrutiny; policing informed by AV-derived data risks reproducing systemic discrimination under a veneer of technological objectivity. The politics of predictive mobility underscore urgent ethical and regulatory dilemmas.

Beneath these developments lies a deeper transformation: movement itself becomes metadata. AV-generated data can reconstruct granular details of daily routines, social interactions, and lifestyle patterns. When aggregated and monetised, this information feeds advertising algorithms, urban planning, and surveillance apparatuses alike. The trade-off between convenience and control becomes strikingly tangible: passengers gain seamless mobility experiences but relinquish intimate behavioural data with broad social consequences. Questions of autonomy, consent, and the contours of public and private space are no longer abstract—they are in motion, quite literally.

Through case studies ranging from urban AV fleets to rural pilot schemes, this chapter interrogates the entanglement of technology, power, and surveillance in motion. It situates AV data practices within broader debates over digital rights and governance, highlighting emergent frameworks attempting to balance innovation with privacy protection. Navigating this terrain demands an understanding not only of the technical machinery but of the geopolitical stakes and societal consequences embedded in autonomous mobility.

Readers are invited to see AVs as nodes within vast data networks—machines that enable and constrain human movement through largely invisible mechanisms of observation and control. This exploration encourages critical engagement with the promises and perils of surveillance in motion, prompting reflection on what it truly means to move freely when every route, pause, and interaction is recorded, analysed, and potentially exploited.

By examining these dynamics, the chapter advances a vital conversation about autonomy in a world where mobility and data are inseparable. It compels reconsideration of the value of privacy and human agency amid rapid technological transformation, urging vigilance and proactive governance to ensure that the silent chauffeur does not become a conduit for unchecked observation.

Data as Fuel

Autonomous vehicles (AVs) are often portrayed as marvels of engineering and artificial intelligence, yet beneath their sleek exteriors lies a relentless engine powered not by petrol or electricity, but by data. This data—an unceasing stream of sensory input, environmental metrics, and behavioural signals—is the fundamental resource driving every decision an AV makes. To function effectively and safely, AVs depend on continuous surveillance of their surroundings, gathering information not only about roads and traffic but also about passengers, pedestrians, and the urban environment. This pervasive data dependency raises profound questions about consent, privacy, and the broader social implications of living in a world where movement itself is under constant observation.

The operational demands of autonomous vehicles necessitate a level of environmental awareness far surpassing human capability. Equipped with an array of sensors—cameras, lidar, radar, ultrasonic devices—and interconnected through vehicle-to-everything (V2X) communication networks, AVs continuously scan and interpret a complex web of variables. These include lane markings, traffic signals, the behaviour of other road users, weather conditions, and even subtle cues such as road surface texture or ambient light. The resulting data feed is enormous and persistent, requiring sophisticated algorithms to process and translate into real-time navigation decisions. *In essence, data serves as the lifeblood of autonomy, animating the vehicle's capacity to perceive and respond.*

Yet the collection of such extensive data inevitably encroaches on privacy, challenging traditional understandings of informed consent. Unlike static surveillance systems, AVs move through private and public spaces alike, capturing details that extend beyond mere operational necessity. Passengers' conversations, biometric identifiers such as gait or facial recognition, and behavioural patterns may all be recorded, transmitted, and stored. The boundary between what is necessary for safe navigation and what constitutes intrusive monitoring is blurred, with users often unaware of the scope and fate of their data. *Consent, in this context, is complicated by the opacity of data practices and the asymmetry of knowledge between technology providers and users.*

Furthermore, the interconnected nature of AVs amplifies these concerns. Vehicles communicate with infrastructure, other vehicles, cloud platforms, and service providers, generating a distributed data ecosystem. This networked architecture facilitates not only navigation but also the aggregation of vast datasets used for traffic optimisation, predictive maintenance, and service customisation. However, it also creates multiple points of vulnerability, where data can be intercepted, misused, or commodified without explicit user approval. For example, ride data might be sold to advertisers or leveraged by insurance companies to tailor premiums, often without clear transparency. *Such practices highlight the tension between the functional necessity of data collection and the ethical imperative to protect individual autonomy.*

Real-world applications underscore these challenges. Early AV pilots in cities such as Phoenix and San Francisco have demonstrated how constant data collection shapes user experience and operational oversight. In some cases, passengers have reported discomfort with the sense of being monitored not only by the vehicle's cameras but by the broader surveillance ecosystem. Legal frameworks in many jurisdictions remain underdeveloped, struggling to keep pace with the speed and complexity of AV data flows. Existing data protection laws, crafted for more static digital interactions, often fail to account for the mobility and granularity of AV-generated data. *This regulatory lag exacerbates uncertainty and erodes public trust.*

Moreover, the imperative for real-time data processing necessitates that much information be transmitted and analysed off-vehicle, frequently in cloud environments controlled by private corporations. This model raises additional concerns about data sovereignty, jurisdiction, and corporate accountability. Who owns the data generated during a ride? How long is it retained, and for what purposes? Can users demand deletion or restrict sharing? These questions remain largely unresolved, placing individuals in a precarious position regarding their digital rights. *The commodification of mobility data creates power asymmetries that extend beyond transport, influencing surveillance capitalism and state security regimes.*

The philosophical dimensions of this data-centric model of mobility demand scrutiny. Autonomous vehicles transform movement into a continuous flow of data points subject to interpretation and intervention. The lived experience of travel becomes inseparable from digital traceability, raising fundamental questions about freedom, autonomy, and control in public and private spaces. *When every turn, stop, and interaction is quantified, how do notions of anonymity and spontaneity persist?* This convergence of mobility and surveillance reshapes urban life and social relations, embedding technology-mediated observation into the fabric of daily existence.

Technological responses to privacy concerns—such as data anonymisation, encryption, and user consent protocols—offer some mitigation but are not panaceas. The scale and granularity of AV data challenge even the most advanced privacy-preserving techniques. Furthermore, the commercial incentives to monetise data create conflicting pressures that complicate ethical data governance. Efforts to develop transparent and accountable frameworks must grapple with the intricate interplay of technological feasibility, economic interests, and democratic values. *This necessitates multi-stakeholder engagement involving technologists, policymakers, civil society, and users themselves.*

In sum, the reliance of autonomous vehicles on constant data surveillance constitutes a fundamental transformation in mobility—

one that intertwines functionality with complex ethical, legal, and social issues. Understanding data as the essential fuel for AV operation reveals the necessity of reimagining consent, privacy, and power in this new landscape. For those navigating the promise and peril of the silent chauffeur, grappling with the implications of data as fuel is indispensable to shaping a future where technology serves human dignity rather than diminishes it.

The Mapping Arms Race

The foundation of autonomous vehicle (AV) technology rests upon the bedrock of real-time, high-precision mapping. Yet these digital cartographies are far more than navigational aids—they have become geopolitical instruments in a rapidly intensifying competition among states and corporations. The struggle to dominate this mapping landscape is shaping global power dynamics in subtle but profound ways, as control over the data that charts our roads translates into influence over the future of mobility, urban governance, and national security. Understanding the contours of this mapping arms race is essential to grasping the broader implications of AV deployment worldwide.

Unlike traditional maps designed for human interpretation, AV maps are living datasets—constantly updated, extremely detailed, and integrated with sensor networks that feed real-time information about traffic flows, road conditions, pedestrian movement, and infrastructure changes. These maps must function with centimetre-level accuracy and split-second responsiveness to enable vehicles to make safe decisions. Creating, maintaining, and controlling such maps demands immense computational resources, proprietary algorithms, and a continuous influx of data from millions of vehicles and infrastructure sensors. *This technological complexity has transformed mapping from a benign geographic exercise into a strategic asset.*

Governments have taken note of the strategic value embedded in digital mapping. Control over geospatial data offers potential leverage in military logistics, intelligence gathering, and cyber operations. In the United States, agencies like the Department of Defense and National Geospatial-Intelligence Agency have long recognised the importance of spatial data dominance. This understanding now extends to the civilian sphere, where federal investments support initiatives to develop national AV infrastructure and secure domestic control over mapping technologies. Conversely, China pursues an assertive policy of technological self-reliance, investing heavily in homegrown mapping platforms and integrating them into its expansive surveillance and smart city projects. *These competing approaches reflect divergent political models and strategic imperatives, transforming maps into symbols of sovereignty and technological autonomy.*

Private corporations—particularly tech giants and specialised mapping firms—are pivotal actors in this contest. Companies like Google's Waymo, Apple, Tesla, and Chinese firms such as Baidu and Alibaba have amassed proprietary mapping databases enriched by the data collected from their autonomous and connected vehicles. These entities wield significant market power, as access to up-to-date, high-fidelity maps is a prerequisite for AV functionality. The commercial stakes are enormous: mapping data underpins not only navigation but also urban planning, logistics optimisation, insurance risk assessment, and advertising targeting. The desire to protect intellectual property and competitive advantage often leads to restrictive licensing and limited data sharing, exacerbating tensions around transparency and equitable access. *Here, commercial interests intersect with geopolitical competition, complicating governance.*

The competitive landscape is further complicated by cross-border data flows and regulatory regimes. Some countries impose strict data localisation laws, mandating that geospatial data generated within their borders remain under domestic control. Such regulations aim to safeguard national security but can fragment the global mapping ecosystem, hindering interoperability and slowing technological diffusion. International cooperation is limited, in part due to mistrust and differing legal frameworks concerning data privacy, ownership, and security. *The absence of robust multilateral governance creates a patchwork of standards that shape how, where, and by whom mapping technologies are deployed.*

Case studies illustrate the stakes involved. Singapore's development of a comprehensive, government-controlled digital twin of the city exemplifies a state-led approach to mapping, integrating real-time data streams to optimise urban mobility and enhance security.

Conversely, in the United States, the dominance of private mapping platforms creates a reliance on commercial entities for critical infrastructure data, raising concerns about monopolistic control and resilience. In Europe, initiatives such as the European Location Framework seek to promote data sharing and interoperability to reduce dependency on foreign platforms. These contrasting models highlight how governance philosophies and market structures influence the mapping arms race.

Beyond geopolitics and commerce, the mapping arms race has profound social and ethical dimensions. The granular, real-time nature of AV maps enables pervasive surveillance capabilities that can erode privacy and civil liberties. The integration of facial recognition, behavioural analytics, and predictive policing tools with mapping data raises alarms about the expansion of state and corporate monitoring. Furthermore, the prioritisation of mapping technologies for affluent urban centres risks exacerbating spatial inequalities, as under-mapped regions may become excluded from the benefits of autonomous mobility. *Thus, the contest for mapping dominance also shapes whose voices and places are rendered visible or invisible.*

The technological challenges of maintaining real-time, high-precision maps demand constant innovation. AVs must reconcile static map data with dynamic environments—temporary roadworks, accidents, or unexpected obstacles—necessitating sophisticated sensor fusion and machine learning. Ensuring map accuracy and freshness requires continuous data collection and validation, often through fleet-based mapping vehicles or crowdsourcing from users. However, the reliance on such distributed data gathering raises vulnerabilities to cyberattacks and data manipulation, threatening the safety and reliability of autonomous systems. *In this context, mapping is not merely a technical problem but a complex socio-technical enterprise requiring robust security and governance.*

The mapping arms race also intersects with broader debates about data governance and digital sovereignty. Questions of who owns, controls, and profits from mapping data are central to discussions about the equitable distribution of technological benefits. Proposals for open data initiatives and public mapping infrastructure seek to democratise access and reduce monopolistic barriers, but face resistance from entrenched commercial interests. The balance between innovation incentives and public accountability remains delicate, with implications for the future shape of smart cities and connected mobility ecosystems.

In sum, the competition to control real-time, high-precision maps encapsulates a crucial dimension of the autonomous vehicle revolution—one where technology, geopolitics, commerce, and social justice converge. Mapping is no longer a passive backdrop but an active determinant of mobility, security, and power. As AVs become increasingly embedded in daily life, the outcomes of this arms race will influence not only how we move but who controls the very fabric of urban space and information. Understanding this contest equips readers to engage critically with the promises and perils of autonomous mobility in a digitally mapped world.

Predicting You, Predicting Risk

The transformation of autonomous vehicles (AVs) from mechanical marvels into omnipresent agents of mobility depends not only on advanced sensors and artificial intelligence but fundamentally on the relentless harvesting and analysis of behavioural data. Every turn taken, every route optimised, and every pause registered contributes to an expansive, dynamic profile of individual and collective movement patterns. This data-driven predictive capacity powers more than just navigation—it increasingly underwrites decisions in insurance, law enforcement, and access to services, turning the very act of movement into a quantifiable risk metric. Such developments raise profound questions about autonomy, privacy, and justice in a world where algorithms pre-emptively judge individuals based on the footprints they leave on the road.

Behavioural data from AVs extends far beyond traditional notions of driving behaviour. It encompasses not only speed and braking habits but also patterns of route selection, time of travel, interaction with other vehicles and pedestrians, and even biometric or environmental cues detected by sensors. This granular data enables the construction of sophisticated risk profiles, which insurers now use to dynamically adjust premiums, sometimes in real-time. Early experiments in usage-based insurance (UBI) harness telematics to reward safer driving with lower costs; yet as AVs take over the wheel, this model evolves into a complex, algorithmic appraisal of "predictive risk." *While ostensibly designed to improve safety and efficiency, this system can embed and amplify existing social inequalities.* For example, individuals living in neighbourhoods with higher accident rates or traffic enforcement may be assigned higher risk scores, independent of their personal behaviour, leading to disproportionate insurance costs or restricted access to AV services.

Policing and public safety agencies also leverage behavioural data harvested from AVs, often in less transparent ways. Predictive policing algorithms, which have been widely criticised for reinforcing systemic biases, find new dimensions when integrated with real-time mobility data. Routes frequently travelled, durations spent in particular areas, and social interactions detected through sensor networks feed into risk assessments that influence patrol patterns, stops, and surveillance priorities. The integration of AV data with law enforcement databases raises the spectre of constant monitoring under the guise of traffic safety, potentially eroding civil liberties. The data-driven logic can justify intensified scrutiny of particular communities, thus entwining the promise of technological progress with the persistence of social control. *The deployment of AVs inadvertently extends the reach of surveillance apparatuses into the fabric of everyday movement.*

Access to autonomous mobility services becomes another terrain where behavioural data exerts power. Service providers use predictive algorithms to allocate resources dynamically, adjusting availability and pricing based on user profiles and perceived risk. This stratification may limit AV access for individuals flagged as "high risk," a designation derived from patterns shaped by socioeconomic factors beyond their control. Consequently, the AV revolution risks deepening mobility inequities, as those most dependent on affordable transportation face barriers rooted in opaque algorithmic decisions. *Rather than a universal benefit, autonomous mobility may replicate and entrench existing divides in who can move freely and safely.*

The ethical implications of predictive behavioural data extend to questions of consent and transparency. Users often have limited understanding or control over the extent and use of their data, raising concerns about informed consent in an environment where opting out could mean losing access to essential services. The proprietary nature of many AV data systems compounds this opacity, as commercial interests prioritise data monetisation and competitive advantage over user rights. Philosophically, this challenges traditional conceptions of agency and responsibility: if risk and opportunity are increasingly determined by algorithmic predictions rather than human choice, what does autonomy mean in mobility? *The subtle shift from active driver to passive data point redefines the relationship between individuals and their movements.*

Real-world cases illustrate these dynamics. For instance, early pilots of AV taxi services in metropolitan areas have reported instances where users with certain behavioural profiles faced higher surcharges or reduced service options. Insurance companies experimenting with AV telematics have encountered controversies over the fairness of automated premium adjustments based on data interpreted without human context. Law enforcement collaborations with tech firms have sparked debates about data sharing agreements and the potential for function creep from traffic safety to broader surveillance. Each example underscores the complex interplay between data, prediction, and societal values.

Technically, the challenge lies in the design of algorithms that balance safety and fairness, ensuring that risk assessments do not become proxies for discrimination. Machine learning models trained on historical data may inherit and perpetuate biases present in that data, necessitating rigorous auditing and transparent governance. Moreover, the scale and velocity of AV data require new regulatory frameworks that can protect individual rights without stifling innovation. Internationally, approaches vary widely, with some jurisdictions imposing strict data protection laws and others favouring market-driven experimentation, creating a patchwork of protections and risks.

As behavioural data becomes the fuel driving AV ecosystems, the stakes extend beyond the technical into the social contract between individuals, corporations, and the state. Questions of who sets the parameters for risk, how accountability is maintained, and how benefits and harms are distributed are not merely policy concerns but moral imperatives. Engaging with these issues demands a multidisciplinary perspective, blending data science, ethics, law, and urban planning to shape a mobility future that respects human dignity and promotes equitable access.

Ultimately, the power to predict risk from behavioural data in the AV era is a double-edged sword: it offers the potential to enhance safety and efficiency but risks entrenching inequality and surveillance under the veneer of objectivity. The choices made now in governance, technology design, and public discourse will determine whether predictive mobility becomes a tool of empowerment or exclusion. For readers committed to understanding and shaping this transformative moment, unpacking the nuances of behavioural data's role in predicting risk is essential to envisioning a just autonomous future.

When Movement Becomes Metadata

The act of moving through a city, once an inherently private and unquantified experience, is rapidly being transformed into an intricate mosaic of metadata—digitised traces that document not only where individuals go, but how, when, and with whom. Autonomous vehicles (AVs), at the forefront of this transformation, operate as constant sensors and data collectors, converting every journey into streams of monetisable information. This commodification of mobility data reshapes not only transport systems but also the social and economic dynamics of urban life, raising urgent questions about surveillance, consent, and the value of everyday movement.

The technical architecture of AVs necessitates an unprecedented scale of data collection. Unlike human drivers, who navigate based on intuition and local knowledge, AVs rely on continuous inputs from cameras, lidar, radar, GPS, and a suite of environmental sensors to interpret their surroundings. Each trip generates a layered dataset that includes speed, trajectory, passenger behaviour, environmental conditions, and interactions with other vehicles and infrastructure. This data is aggregated and analysed in real time, enabling AVs to make split-second decisions and optimise routes for efficiency and safety. Yet beyond the immediate operational purpose, these datasets become valuable commodities, sought after by manufacturers, service providers, advertisers, insurers, urban planners, and even governments.

The economic imperative to monetise movement metadata drives AV companies to forge complex data ecosystems. User data is sold or shared with third parties, creating revenue streams that often dwarf profits from the vehicles themselves. For instance, ride-hailing services that incorporate AV technology transform mobility into a subscription-like model, where user preferences, travel habits, and demographic profiles inform targeted advertising, dynamic pricing, and market segmentation. This parallels the broader digital economy's reliance on personal data, but with a critical distinction: mobility data is uniquely sensitive, revealing intimate patterns of daily life that can be traced back to homes, workplaces, social networks, and private routines.

This transformation complicates traditional notions of privacy. While consumers may consent to data collection as part of service agreements, the pervasive and granular nature of AV metadata surpasses what most individuals comprehend or anticipate. The aggregation of seemingly innocuous data points enables detailed behavioural profiling, risk assessment, and even predictive analytics about future movement. Such capabilities underpin surveillance frameworks that are both corporate and state-led, blurring the lines between commercial exploitation and public governance. *The movement of bodies through urban spaces becomes an open ledger accessible to actors far beyond the individual.*

Notably, the monetisation of movement metadata risks reinforcing existing social inequalities. Data-driven services often privilege users with stable digital identities, access to technology, and economic means, while marginalising populations in transport deserts, low-income neighbourhoods, or informal economies. The commodification of movement can lead to "data deserts" where the lack of digital footprints translates into reduced service availability or inflated costs. Moreover, the opacity of data algorithms and the absence of meaningful user control foster asymmetrical power relations, where corporations determine which behaviours are rewarded, penalised, or made invisible.

Beyond commercial dynamics, the transformation of movement into metadata has profound implications for urban governance and civil liberties. Governments increasingly rely on data collected from AVs and related smart infrastructure to manage traffic flows, enforce regulations, and enhance public safety. While these applications promise efficiency gains, they also enable pervasive monitoring and control, potentially normalising a regime of constant observation. The integration of AV movement data with other surveillance modalities—facial recognition, mobile tracking, social media analysis—creates a comprehensive portrait of individual and collective behaviour. This challenges democratic principles of privacy, due process, and consent in the urban context.

Philosophically, the conversion of movement into metadata invites reflection on the commodification of lived experience and the erosion of embodied autonomy. Where once the city was a space for spontaneous encounters, serendipitous detours, and unquantified presence, it increasingly becomes a domain mapped, measured, and monetised. The act of mobility shifts from a personal expression of freedom to a data transaction, shaping not only how people move but how they understand their own agency within the urban environment. *This transformation demands a rethinking of rights and responsibilities in a datafied city.*

Several real-world case studies illuminate these dynamics. In cities like San Francisco and Singapore, pilot AV programmes are accompanied by extensive data-sharing agreements between private companies and municipal authorities, raising debates over transparency and accountability. Meanwhile, insurance firms use AV-generated movement data to introduce nuanced pricing models that reward safe patterns but may penalise vulnerable groups disproportionately. Activist groups have begun to challenge the expansion of surveillance infrastructure tied to AV rollouts, demanding greater public oversight and data sovereignty.

The technical and ethical challenges of managing movement metadata call for novel governance frameworks. Data minimisation principles, user-centric consent models, and transparent algorithmic auditing are emerging as necessary safeguards. Internationally, regulatory responses vary widely, from the European Union's stringent data protection laws to more laissez-faire approaches elsewhere, reflecting different societal priorities and political cultures. The complexity of AV data ecosystems necessitates cross-sector collaboration among technologists, policymakers, civil society, and the public to ensure equitable and accountable data practices.

Ultimately, the transformation of daily movement into monetised, traceable metadata signifies a pivotal shift in how cities operate and how individuals relate to their environment. The promises of efficiency, convenience, and innovation brought by AVs must be balanced against the risks of surveillance, commodification, and exclusion. For those engaging critically with the autonomous future, understanding the implications of movement as data is essential to shaping cities that are not only smart but also just and humane.

Chapter 7
Behavioral Drift

The emergence of autonomous vehicles signals more than a technological milestone; it ushers in a subtle but profound recalibration of human behaviour in relation to machine logic. As these systems assume increasing control over the mechanics of movement, we begin to adjust—often unconsciously—our routines, perceptual habits, and cognitive engagement with the act of travel. This chapter explores what might be called behavioural drift: the gradual reshaping of human agency in response to the affordances and constraints of automation, and the implications this holds for spontaneity, awareness, and trust.

At the heart of this transformation lies a crucial insight: machines do not simply respond to human preferences—they actively mould them. Algorithms that optimise for efficiency or safety silently suppress the improvisational shortcuts that once reflected local knowledge or serendipitous discovery. Urban spaces, historically navigated through a tapestry of personal choices, risk being reconstituted as predictable grids governed by data imperatives. The disappearance of the shortcut is emblematic of a broader erosion of autonomy, where individual ingenuity and intimate familiarity with the city yield to the uniformity of algorithmic convenience.

The impact extends beyond the physical act of movement. Delegation fatigue emerges as people shift from active decision-makers to passive passengers, ceding control to systems that promise ease. The mental work inherent in driving—judging hazards, anticipating behaviour, making split-second decisions—gradually diminishes. Over time, situational awareness dulls, cognitive agility wanes, and the ability to intervene effectively when automation falters becomes uncertain. What was once routine skill now risks atrophy under the seduction of convenience.

Trust, too, is reconfigured in this new landscape. Overtrust—an unearned confidence in the infallibility of machines—can amplify rather than mitigate risk. This phenomenon arises from the apparent competence of AV systems, the user's lack of familiarity with technological limitations, and social narratives that valorise automation. The resulting dynamic is precarious: human oversight remains essential, yet confidence and complacency can erode vigilance, leaving a fragile balance between reliance and control.

Empirical observations from pilot deployments and interactions with driver-assist technologies reveal the tangible contours of behavioural drift. Users adjust to the perceived needs of the machine, adhering rigidly to speed limits, avoiding manoeuvres the system finds challenging, and gradually relinquishing the discretionary judgements that once defined their role. This co-evolution of human and machine illustrates adaptation in practice—but also surfaces critical questions about agency, identity, and freedom in an environment increasingly structured by algorithms.

Behavioural drift also resonates with broader cultural and philosophical concerns. The automation of driving forces a reconsideration of autonomy and responsibility: what becomes of the tacit knowledge, embodied skills, and ethical judgements that have long characterised human drivers? The shift invites reflection not merely on technical integration, but on the psychological and societal consequences of delegating everyday decision-making to machines.

For professionals, policymakers, and scholars navigating the autonomous future, behavioural drift demands careful attention. It underscores the importance of designing systems that sustain cognitive engagement, calibrate trust appropriately, and preserve opportunities for improvisation and local expertise. This chapter offers a framework to grapple with these challenges, blending critical insight with practical guidance to support a more balanced, human-centred approach to mobility in an era of automation.

How Machines Shape Us Back

The arrival of autonomous vehicles (AVs) signals a profound recalibration not only in technology but in human behaviour itself. As these sophisticated systems assume control over decisions once firmly in the human domain, a subtle yet significant phenomenon emerges: humans begin to adapt their behaviours to fit the parameters set by machines. This adaptation, far from a passive surrender, represents a complex negotiation where human agency is redefined through interaction with machine expectations. It challenges conventional notions of autonomy, highlighting how technology reshapes human action and cognition in ways both visible and invisible.

The fundamental insight is that automation is not simply a tool that humans wield; it is an active participant in shaping behaviour. Early research in human-machine interaction reveals that users often modify their driving styles, routines, and risk assessments to align with the operational norms embedded within AV systems. For example, drivers operating semi-autonomous vehicles may adhere more rigidly to speed limits or traffic regulations, aware that deviations could confuse or override the machine's algorithms. This adaptation manifests as a *machine-shaped conformity*, where the unpredictability and creativity that once characterised human navigation give way to a more homogenised mode of conduct designed to accommodate the system's logic.

This shift extends beyond individual behaviours to collective social practices. Urban mobility, traditionally informed by a patchwork of local knowledge, social conventions, and informal adaptations, faces transformation as AV networks prioritise standardisation and predictability. In cities piloting AV services, residents have reported changes in pedestrian crossing habits, route choices, and even the timing of travel, influenced by the predictable yet inflexible patterns programmed into these vehicles. Such feedback loops reveal how the machine's design parameters ripple outward, reshaping not only how people drive but also how they move through and perceive urban spaces.

One notable consequence is the erosion of *tacit knowledge*—the intuitive, experience-based understanding of driving that shapes decision-making in complex environments. As machines increasingly dictate acceptable behaviours, drivers and pedestrians alike may lose familiarity with nuanced spatial cues and improvisational skills. This effect is compounded by the *delegation* of responsibility to AV systems, where the human role shifts from active agent to passive monitor. Over time, this can diminish situational awareness and cognitive engagement, fostering a behavioural atrophy that undermines both safety and personal agency.

Yet, this dynamic is far from uniform or uncontested. Variations in machine design, cultural contexts, and regulatory frameworks introduce tensions and contradictions in how humans adapt. In regions where AVs operate alongside traditional vehicles, for example, human drivers often exhibit resistance or creative circumvention of machine norms, asserting control in subtle acts of defiance. Conversely, in highly regulated environments with extensive AV deployment, the behavioural shifts tend toward conformity, raising questions about freedom, diversity, and equity in mobility practices.

Moreover, the interplay between humans and machines invites reflection on deeper philosophical questions about autonomy and control. If humans alter their behaviour to fit machine expectations, to what extent do they remain autonomous actors? Does adaptation to algorithmic governance represent empowerment through collaboration, or a loss of self-determination to technological imperatives? These questions resonate beyond transportation, touching on broader societal debates about the role of automation in shaping human identity and social order.

Real-world case studies illuminate these dynamics with clarity. In Phoenix, Arizona, where autonomous ride-hailing services have operated for several years, researchers observed passengers adjusting their boarding times and destinations to align with AV availability and operational constraints. Similarly, truck drivers transitioning to automated freight systems often report adapting their routines and communication styles to accommodate machine interfaces and scheduling algorithms. These adaptations, while practical, signify a profound behavioural reorientation shaped by the machine's parameters.

Importantly, the reciprocal shaping of human behaviour by machines also carries implications for design and policy. Recognising that AVs influence user conduct underscores the need for system designs that respect and preserve human variability, spontaneity, and judgment. Human-centred design approaches advocate for interfaces and algorithms that accommodate diverse behaviours and provide transparent feedback, mitigating the risk of behavioural narrowing. Policy frameworks must similarly address the social dimensions of adaptation, ensuring that the diffusion of AVs does not erode human agency or exacerbate inequalities.

In sum, the story of autonomous vehicles is not simply one of technological innovation but a co-evolutionary narrative where machines and humans shape each other. The behavioural modifications that humans undertake to meet machine expectations reveal a layered and dynamic relationship, challenging simplistic notions of control and autonomy. This chapter offers a nuanced examination of that relationship, exploring the ways in which human behaviour is moulded by—and, in turn, moulds—the machine logic that increasingly governs our mobility. Understanding this mutual shaping is essential for forging a future where technology enhances rather than diminishes human freedom and dignity.

The Death of the Shortcut

The allure of the shortcut—a quick, unconventional path that shaves minutes off a journey—has long been a hallmark of human navigation. From the informal back alleys known only to locals to the instinctive detours forged by drivers' intimate knowledge of traffic patterns, shortcuts embody a form of *tacit knowledge* that blends experience, intuition, and spontaneity. Yet, in the emerging landscape dominated by autonomous vehicles (AVs), this once-vibrant facet of mobility is steadily fading. Route optimisation algorithms, prized for efficiency and predictability, systematically exclude the nuanced local knowledge and impromptu decisions that define human driving. The death of the shortcut signals a profound transformation—not just in how we move, but in how we engage with our environments, understand space, and exercise autonomy.

At the core of this shift lies the technological imperative for precision and predictability. AVs rely on sophisticated algorithms designed to select routes that optimise for travel time, safety, and energy consumption. These algorithms synthesise data streams—

from real-time traffic to detailed digital maps—then calculate the most efficient path in milliseconds. While the benefits of such optimisation are clear in terms of reduced congestion and emissions, the cost is an inherent rigidity. Unlike human drivers who might seize upon a narrow street, a seldom-used cut-through, or a momentary opportunity to circumvent traffic, AVs operate within pre-defined parameters that privilege *standardised* routes over personalised shortcuts.

This transformation has tangible implications for urban mobility. Consider the everyday experience of navigating a city like London or New York, where local drivers often develop an instinctive sense of less-trafficked streets, one-way systems, or temporary diversions. These informal strategies not only save time but foster a nuanced spatial awareness and engagement with the city's rhythms. When AVs supplant human judgment with algorithmic routing, these forms of *local knowledge* become redundant. The result is a homogenisation of movement patterns, potentially reinforcing congestion on main arteries while underutilising secondary roads. This dynamic runs counter to traditional notions of mobility as a form of personal freedom and spatial mastery.

The disappearance of shortcuts also curtails spontaneity, a subtle yet profound loss. Human drivers frequently make real-time decisions influenced by mood, curiosity, or social interaction—detouring to a favourite café, exploring a new neighbourhood, or simply altering plans on a whim. Automated route optimisation, by contrast, is relentlessly goal-oriented, prioritising *predictability* over flexibility. As a consequence, the rich tapestry of urban exploration risks being reduced to pre-calculated efficiency, stripping journeys of serendipity and personal expression. This raises broader questions about the role of technology in shaping not just movement, but lived experience.

Empirical studies lend weight to these concerns. Research conducted in cities trialling AV services reveals a marked reduction in route variation compared to human-driven vehicles. For instance, a 2023 study analysing ride patterns in Singapore's AV pilot programme found that algorithmically determined routes exhibited 30% less spatial diversity than human drivers' routes over comparable journeys. Passengers reported a diminished sense of connection to their environment, noting that the AV's path felt "mechanical" and "predictable," lacking the nuanced choices their own driving might have offered. Such feedback highlights an emerging tension between technological optimisation and human experiential richness.

Beyond urban form and experience, the death of the shortcut carries implications for resilience and adaptability. Human drivers' ability to exploit shortcuts often serves as a buffer during disruptions—road closures, accidents, or sudden congestion—enabling dynamic responses that maintain mobility. AV route optimisation, constrained by algorithmic logic and digital map fidelity, may struggle to replicate such improvisational agility. Instances of AVs becoming "trapped" in gridlocks due to adherence to rigid routing protocols have been documented, underscoring vulnerabilities inherent in fully optimised but inflexible systems. This challenges claims that AVs will unequivocally enhance urban mobility resilience.

Philosophically, the elimination of shortcuts invites reflection on autonomy and control. Human navigation embodies a form of *situated agency*, where decision-making is informed by embodied experience and contextual judgement. The mechanisation of route selection through AVs shifts this agency from humans to algorithms, raising ethical and existential questions about the *loss of mastery* over our spatial environment. Is mobility reduced to compliance with machine logic, or can new forms of collaborative agency emerge that reconcile optimisation with human values? This tension lies at the heart of debates on the integration of automation in daily life.

Furthermore, the death of the shortcut has uneven social consequences. Access to local knowledge and the ability to improvise routes has traditionally served as a form of social capital, conferring advantages to those embedded in communities. As AVs standardise route selection, these advantages may erode, levelling the playing field in some respects but also marginalising those who rely on informal networks and adaptive skills. Conversely, those with resources to customise AV algorithms or influence infrastructure design could reassert spatial privileges, potentially exacerbating inequalities. Understanding these dynamics is critical for equitable policy and urban planning.

Mitigating the loss of spontaneity and local knowledge requires deliberate design choices. Emerging research advocates for *hybrid routing models* that incorporate stochastic elements, allowing AVs to introduce controlled variability and adapt to local conditions. Integrating community feedback and participatory mapping initiatives can also enrich algorithmic data with lived experience, preserving some aspects of tacit knowledge. Additionally, fostering human-machine interfaces that enable passengers to influence route choices could restore a measure of agency and unpredictability to journeys. These approaches underscore that the death of the shortcut is not inevitable but contingent on technological and governance decisions.

In conclusion, the demise of the shortcut is emblematic of broader shifts accompanying autonomous vehicle integration—shifts that extend beyond efficiency metrics to challenge how we navigate, experience, and relate to urban spaces. As AVs recalibrate the geometry of our streets and the rhythms of our movement, the disappearance of local knowledge and spontaneity calls for critical reflection. It urges stakeholders to balance the undeniable gains of optimisation with the preservation of human creativity, adaptability, and autonomy. Only by acknowledging and addressing these subtle but profound transformations can we ensure that the future of mobility remains not only efficient but also richly human.

Delegation Fatigue

The promise of autonomous vehicles (AVs) lies not merely in their ability to navigate roads independently but in their capacity to free humans from the cognitive burden of driving. Yet, this liberation is paradoxically entwined with a subtle erosion of human attentiveness and cognitive agility—a phenomenon increasingly recognised as *delegation fatigue*. As control is ceded to machines, human users risk becoming passive spectators, their awareness dulled by reliance on automation. This trade-off between convenience and cognitive engagement carries profound implications, not only for road safety but also for broader questions about agency, skill retention, and the evolving human-machine relationship.

Fundamentally, delegation fatigue emerges from the interplay of psychological and physiological factors inherent to human cognition. The act of driving engages a complex array of perceptual, motor, and decision-making faculties, requiring continuous situational awareness and rapid problem-solving. When these responsibilities are handed over to AV systems, drivers enter a state of reduced vigilance, sometimes termed *automation complacency*, where their role shifts from active operator to passive monitor. This shift diminishes mental stimulation, leading to what cognitive scientists describe as "vigilance decrement"—a progressive decline in attention over time during tasks requiring sustained monitoring.

Empirical research vividly illustrates the risks posed by this decline. Studies conducted in driving simulators reveal that participants supervising highly automated vehicles experience slower reaction times and increased error rates when prompted to intervene unexpectedly. In one notable experiment, subjects overseeing semi-autonomous driving for prolonged periods displayed impaired hazard detection compared to manual driving conditions, underscoring how passive roles impair the brain's capacity to maintain readiness. This phenomenon echoes findings from aviation, where pilots overseeing automated systems face similar challenges, further reinforcing the cross-domain nature of delegation fatigue.

The practical consequences on public roads are far from hypothetical. High-profile incidents involving AV prototypes—such as the 2018 fatal crash of a self-driving Uber in Arizona—highlight how diminished human attentiveness can have catastrophic outcomes. Although the vehicle's automation bore significant responsibility, the human safety driver's failure to intervene was also linked to cognitive disengagement attributed to passive monitoring. Such cases underscore a core paradox: while automation aims to enhance safety by reducing human error, it simultaneously introduces new vulnerabilities by weakening human oversight.

Beyond immediate safety concerns, delegation fatigue implicates deeper issues of skill atrophy and identity. Driving is not merely a mechanical task; it forms part of many individuals' sense of autonomy and competence. As automated systems take over, the once-familiar rituals of navigation, judgement, and control risk becoming obsolete. Psychological studies on automation's impact in various industries reveal that prolonged reliance on machines can erode critical skills, diminish confidence, and provoke feelings of helplessness or alienation. The loss of driving proficiency is particularly consequential because it entails not only technical skills but also spatial cognition and decision-making shaped by environmental interaction.

Moreover, the social and cultural dimensions of delegation fatigue merit careful consideration. In societies where driving is a rite of passage or a symbol of independence, surrendering control to AVs challenges entrenched narratives about mobility and freedom. The diminished role of human agency risks fostering a dependency culture, wherein individuals may become reluctant or ill-equipped to assume control in emergencies or when automation fails. This dependency also reshapes legal and ethical frameworks, complicating questions of responsibility and accountability in incidents involving mixed human-machine control.

Mitigating delegation fatigue requires nuanced design and policy interventions. From a technological standpoint, AV systems must balance automation with mechanisms that sustain human engagement. Research advocates for *adaptive automation*—systems that modulate the level of control granted to the human operator based on context, workload, and attentional state. For example, systems that intermittently require user input or provide subtle alerts can maintain vigilance without overwhelming the driver. Advances in biometric monitoring, such as eye-tracking and brainwave analysis, promise real-time assessment of driver attentiveness, enabling dynamic adjustments to interface and control.

Education and training also play pivotal roles. Preparing users to understand the capabilities and limitations of AV systems fosters realistic expectations and encourages active supervision. Simulation-based training can cultivate skills necessary for seamless transitions between automated and manual control, reducing the cognitive shock and confusion that contribute to fatigue. Equally important is cultivating a culture of *responsible delegation*, where users recognise their enduring role in safety despite automation's allure.

Policy frameworks must evolve to address delegation fatigue comprehensively. Regulatory standards should mandate performance benchmarks not only for technical reliability but also for human-machine interaction quality. Guidelines on monitoring driver engagement and mandatory breaks during prolonged automation use could parallel existing regulations in commercial driving and aviation. Furthermore, transparency in AV capabilities and limitations must be enforced to prevent overtrust—another facet intertwined with delegation fatigue—ensuring users do not overestimate system autonomy and underestimate their ongoing responsibilities.

Philosophically, delegation fatigue challenges prevailing techno-optimist narratives that present automation as unequivocal progress. It invites a reassessment of human-machine collaboration, urging a recognition that automation reconfigures rather than eliminates human roles. The question shifts from *how* to automate to *how* to automate *with* humans as active partners, preserving cognitive vitality and agency. This reframing aligns with broader debates on technology's impact on human skills, urging a balance between efficiency gains and the preservation of human capacities essential to autonomy and safety.

In sum, delegation fatigue illuminates the complex cognitive and social dynamics that accompany the transfer of control to autonomous systems. It highlights the necessity of designing AV technologies and policies that sustain human engagement rather than render it obsolete. As vehicles become smarter, so too must our understanding of human psychology and responsibility deepen, ensuring that the promise of automation does not come at the cost of vigilance, skill, and ultimately, human dignity on the road.

Trust and Overtrust

Trust forms the invisible infrastructure upon which the entire edifice of autonomous vehicle (AV) technology depends. Yet, as users increasingly place their confidence in these systems, a disquieting tension emerges between justified trust and its overextension—a phenomenon known as *overtrust.* The seductive allure of automation's promise can breed an unearned confidence, obscuring the inherent limitations and occasional fallibility of AV systems. This imbalance carries significant implications, magnifying risk not only for individual users but for society at large. Understanding the nuanced dynamics of trust and overtrust is essential to navigating the complex psychological and practical challenges presented by the integration of AVs into daily life.

At its core, trust in AVs reflects a rational response to demonstrated reliability and anticipated benefits. When technology consistently performs as expected—navigating complex traffic conditions, responding adeptly to hazards, reducing human error—users develop confidence that delegation is safe. This trust is critical, for without it, the potential of AVs to transform mobility would falter. However, *trust is not a binary attribute but a spectrum*, contingent on experience, transparency, and understanding. It is here that the paradox of overtrust arises: trust that exceeds the actual capabilities and design limitations of the technology.

The roots of overtrust are multifaceted. Human cognitive biases, such as the automation bias—where users tend to favour suggestions from automated systems over their own judgment—play a pivotal role. This bias is compounded by the *illusion of control* that sophisticated AV interfaces often create. Features that present a veneer of seamless control may lead users to underestimate residual risks or the necessity of vigilant oversight. For example, early iterations of Tesla's Autopilot system demonstrated how marketing language and interface design can contribute to overestimation of system autonomy, resulting in misuse or complacency.

Empirical studies have illustrated the consequences of overtrust in AV contexts. Laboratory experiments and real-world observations reveal that users often reduce their situational awareness, failing to monitor system performance adequately. This complacency manifests in slower reaction times during critical moments requiring human intervention. Notably, the fatal Uber self-driving car incident in 2018 exposed how an overreliance on automation, coupled with diminished human vigilance, can precipitate tragedy. Here, the safety driver's inattentiveness was not merely individual negligence but symptomatic of a broader trust dynamic skewed by overconfidence in machine capability.

Moreover, overtrust carries systemic risks that extend beyond individual accidents. Widespread uncritical acceptance of AV infallibility risks normalising a lowered threshold for oversight, regulatory scrutiny, and accountability. It can foster societal complacency, where emerging failures or vulnerabilities in AV technologies are overlooked or underprioritised until consequences become severe. This dynamic challenges governance frameworks, which must calibrate safety standards and enforcement mechanisms to counterbalance tendencies toward unwarranted trust.

Yet, a delicate balance must be struck. Excessive scepticism or distrust can hinder AV adoption and slow progress toward the considerable safety and efficiency gains these technologies promise. The challenge lies in cultivating *calibrated trust*—a relationship between user and machine that is neither uncritical acceptance nor paralyzing doubt. Achieving this requires transparency in system design, clear communication of capabilities and limitations, and fostering user education that encourages active engagement rather than passive reliance.

Philosophically, the phenomenon of overtrust invites reflection on the evolving nature of human-technology interaction. It underscores the complexity of transferring agency from human to machine, revealing that trust is not simply given but constructed and negotiated within sociotechnical systems. Trust in AVs encapsulates broader questions about how societies delegate critical tasks to algorithms and the ethical imperatives that accompany such delegation.

Technological design plays a crucial role in mediating trust relationships. Emerging approaches advocate for *explainable AI* and *human-centred automation*, which prioritise systems that provide interpretable feedback and invite user understanding. For example, interface elements that convey the machine's confidence level or rationale for specific actions can help users form more accurate mental models, mitigating the risk of overtrust. Similarly, *fail-safe* mechanisms that gracefully degrade automation in uncertain conditions require clear user prompts to re-engage human control, thereby preserving vigilance.

Regulatory responses to overtrust are still nascent but evolving. Policymakers increasingly recognise the necessity of standards that encompass human factors alongside technical specifications. Guidelines mandating continuous monitoring, user training requirements, and clear liability frameworks seek to align user behaviour with safe system operation. Yet, the pace of technological innovation often outstrips regulatory adaptation, leaving gaps where overtrust-related vulnerabilities persist unaddressed.

Contrasting views emerge around the inevitability and manageability of overtrust. Some argue that, like many previous technological revolutions, society will adapt over time, with cultural norms and legal frameworks gradually recalibrating trust dynamics. Others caution that the opacity and complexity of AV systems may entrench overtrust more deeply than past technologies, demanding proactive interventions grounded in interdisciplinary research and policy innovation.

In sum, the dual forces of trust and overtrust in autonomous vehicles illuminate a central paradox in modern technology adoption. While trust is indispensable for leveraging the transformative potential of AVs, unchecked overtrust undermines safety and accountability. Navigating this terrain requires a concerted effort across design, education, regulation, and ethical reflection—ensuring that human confidence aligns appropriately with machine capacity. Only through such calibrated trust can AVs realise their promise without sacrificing the vigilance and agency that underpin safe and responsible mobility.

Chapter 8
Designing for Dependency

The modern fascination with seamlessness promises a life stripped of effort, delay, and friction—a world where technology smooths every wrinkle and anticipates every need. Yet beneath this veneer of convenience lies a complex web of subtle costs: compromises to personal freedom, the shaping of thought, and the quiet erosion of human agency. The ideology of seamlessness is central to autonomous vehicle (AV) systems, influencing not only our movements across physical space but our cognitive and social landscapes. What is presented as ease often conceals mechanisms of behavioural guidance, steering users toward outcomes that serve commercial, institutional, or infrastructural priorities.

This chapter illuminates how AV systems operate not merely as neutral instruments but as active participants in structuring human behaviour. Through the lens of behavioural architecture, we see how software, hardware, and interface design nudge users—sometimes deliberately, more often subtly—toward particular patterns of interaction. These nudges are rarely innocent. They frequently coincide with profit objectives, data collection imperatives, and broader power dynamics that redefine the user–machine relationship. A routing algorithm or interface, seemingly impartial, becomes an invisible arbiter, shaping not just travel choices but attention, trust, and expectation.

Beyond immediate behavioural effects lies a quieter, cognitive hazard: the weakening of human resilience in the face of relentless optimisation. AV systems promise efficiency—fewer delays, smoother commutes, precise navigation—but the very predictability they impose can erode the unpredictability that fosters creativity, improvisation, and situational awareness. Behaviour is channelled into narrow trajectories; spontaneity is marginalised. The friction of decision-making, long a source of skill, judgment, and adaptability, gives way to automated pathways. Convenience, in this sense, comes at the cost of cognitive and social richness.

Perhaps most pressing is the contraction of exit options. As dependence on automated guidance deepens, the capacity to question, override, or disengage from prescribed actions diminishes. Autonomy—once a hallmark of personal dignity—faces systematic constraint. When the manual alternative becomes impractical or effectively impossible, the stakes shift from individual preference to collective norms of agency. Society itself must reckon with what it means to inhabit spaces increasingly designed for machine optimisation rather than human deliberation.

Drawing on empirical studies, case analyses, and philosophical reflection, this chapter urges a critical reassessment of frictionless design. It exposes the trade-offs embedded in AV technologies and their subtle influence on thought, choice, and self-determination. The imperative is clear: design must acknowledge human complexity, safeguarding opportunities for intervention, reflection, and creativity. Only by recognising the interplay between system logic, behavioural influence, and agency can we ensure that autonomous technologies enhance rather than diminish the conditions for human freedom in an increasingly automated world.

Frictionless Isn't Neutral

The pursuit of seamlessness has become a defining characteristic of contemporary technology design, promising effortless interactions, instant gratification, and unprecedented convenience. Yet, beneath this glossy veneer lies a less obvious reality: frictionless experiences are far from neutral. They carry hidden costs—many of which challenge the very notions of freedom, agency, and reflective thought. The ideology of seamlessness, especially within the realm of autonomous vehicles (AVs), not only reshapes how we navigate physical spaces but also subtly reconfigures our cognitive landscapes and social relations.

At first glance, frictionless design appears unquestionably beneficial. Eliminating delays, confusion, and complexity aligns with a broader cultural valorisation of efficiency and productivity. However, the absence of friction often translates into the removal of choice points and moments of deliberation, which are essential to human autonomy. Friction—those minor obstacles, interruptions, and moments of resistance—has traditionally served a vital function: it compels attention, prompts reconsideration, and preserves the space for agency. Without it, users are nudged down predetermined paths with diminished opportunity to pause, question, or deviate. *In designing for frictionlessness, technology engineers a behavioural environment where ease supplants engagement, and control shifts subtly from human to system.*

Autonomous vehicle systems exemplify this phenomenon. Their core selling point—the effortless, hands-off driving experience—demands a high degree of trust in opaque algorithms and predefined routes. The user's role is reduced to passive passenger, ostensibly freed from the mental load of navigation and control. Yet, this liberation comes at the cost of disengagement, eroding the practical knowledge and situational awareness traditionally gained through active driving. What is often overlooked is that the seamless integration of AVs into daily life hinges on constraining user behaviours to those anticipated and permitted by system logic. The frictionlessness of AV travel is, in essence, a carefully curated experience, where deviations can trigger system corrections or even safety interventions. This curated smoothness is not a neutral backdrop but a form of behavioural governance that shapes how individuals move and think.

Moreover, the commercial imperatives underpinning frictionless design complicate its ethical profile. Companies behind AV technologies and related mobility services seek to maximise user engagement and data capture, often leveraging seamlessness as a tool to reduce resistance to surveillance and monetisation. The design of interfaces and interactions nudges users toward behaviours that align with corporate interests, such as preferring particular routes, modes of travel, or services embedded within the AV ecosystem. Such behavioural architecture raises critical questions about consent and the erosion of genuine choice. When seamlessness is engineered to streamline not only movement but also data flows and consumer patterns, the freedom it promises is tethered to underlying economic motivations that users seldom see or fully understand.

The philosophical implications extend further. Seamlessness encourages a mode of existence prioritising efficiency and predictability over reflection and resilience. In a world engineered for speed and ease, the capacity to tolerate uncertainty, to experiment, or to confront the unexpected becomes diminished. The very conditions that nurture creativity, problem-solving, and ethical deliberation risk being undermined by designs that pre-emptively resolve friction. This over-optimisation fosters environments where dependence on automation breeds cognitive complacency and limits the development of adaptive capacities essential for both individual flourishing and democratic life.

Consider the analogy of the autopilot in aviation: it frees pilots from routine tasks but requires vigilance and skill to intervene when systems fail or conditions change. In contrast, frictionless AV systems frequently disallow or discourage meaningful human intervention, leaving users vulnerable to lapses in attention or critical situational judgement. The ideology of seamlessness thus paradoxically engenders new forms of risk, borne not of mechanical failure alone but of diminished human agency.

Critics might argue that seamlessness is simply progress—an inevitable improvement upon clunky, inefficient predecessors—and that any resistance reflects nostalgia or technophobia. Yet, this view underestimates the subtle and pervasive ways design shapes human experience. The trade-offs implicit in frictionless systems demand scrutiny precisely because they transform fundamental aspects of autonomy and social interaction. It is not a question of rejecting convenience but of recognising that *ease is never a neutral good.* Every reduction in friction involves choices about whose interests are prioritised, which behaviours are encouraged, and what forms of engagement are foreclosed.

The path forward requires a reframing of design ethics—one that values the preservation of friction as a space for freedom, reflection, and resilience. This entails embracing complexity rather than smoothing it away, and creating systems that empower users with genuine options rather than seamless submission. In the context of AVs, such an approach might involve transparent interfaces that reveal algorithmic decision-making, flexible modes that allow for human override, and infrastructural designs that preserve the unpredictability and richness of urban mobility.

Ultimately, the critique of frictionless ideology reveals a deeper tension at the heart of contemporary technological culture: the balance between control and autonomy, between efficiency and thoughtfulness. *Frictionless design is a powerful force shaping not only how we move but how we think and who we become.* Understanding and addressing its hidden costs is essential to ensure that the technological futures we build serve human freedom rather than diminish it.

Behavioral Architecture

At the heart of autonomous vehicle (AV) systems lies a subtle but profound form of influence: behavioural architecture. This term refers to the deliberate design of technology interfaces and interactions that guide, shape, and often restrict user behaviour—frequently without explicit consent or awareness. In the context of AVs, this architecture extends beyond mere functionality, embedding within the very mechanics of transport a set of behavioural norms and incentives. The promise of effortless mobility thus comes entangled with a growing reality: users are nudged, sometimes steered, towards specific patterns of action that serve underlying commercial and operational agendas. *Understanding this dynamic is critical not only for grasping how AVs operate but also for appreciating the broader implications for autonomy, agency, and urban life.*

The notion of behavioural architecture builds on decades of insights from behavioural economics and cognitive psychology, which demonstrate that human decisions are rarely made in isolation or through purely rational deliberation. Instead, choices are profoundly shaped by context, presentation, and constraints—factors that designers can manipulate to produce predictable outcomes. AV systems harness these principles with technological precision, integrating data-driven algorithms, real-time feedback loops, and adaptive interfaces that coalesce to form a comprehensive behavioural scaffold. For example, route selection algorithms may prioritise certain pathways not simply because they are fastest but because they optimise fleet efficiency or generate higher revenue through partnerships with commercial zones or service providers. The user's input becomes a variable in a broader optimisation problem, where individual preferences are balanced against corporate or municipal objectives.

Real-world implementations illustrate how these nudges manifest in practice. Consider ride-hailing services deploying AV fleets. The platform may subtly discourage users from requesting pickups in low-demand or less profitable areas by extending wait times or limiting vehicle availability. Conversely, passengers might be incentivised to choose premium services or in-car advertisements through dynamic pricing models or rewards systems. These mechanisms reshape travel habits and consumer behaviour, embedding economic incentives within the ostensibly neutral function of transport. The AV experience is thereby transformed from a passive service into an active site of behavioural modification, where every journey contributes data and revenue streams aligned with specific business models.

Such manipulation raises profound ethical questions. The opacity of algorithmic decision-making compounds concerns, as users often remain unaware of the forces directing their choices or the trade-offs involved. Unlike traditional forms of influence, behavioural architecture in AVs operates at a scale and speed that defy easy scrutiny. The algorithms are proprietary, shielded as commercial secrets, limiting public understanding and accountability. Moreover, the entanglement of behavioural nudges with profit motives risks prioritising commercial optimisation over equitable access, user wellbeing, or even safety. The ethical tension emerges starkly when nudges inadvertently or deliberately exacerbate social inequalities—such as when economically disadvantaged communities receive inferior service quality or are excluded from certain mobility benefits due to profit-driven algorithms.

Critically, the behavioural architecture of AVs does not simply reflect existing social patterns; it actively participates in reshaping them. Over time, repeated exposure to algorithmically guided interactions can recalibrate user expectations, habits, and norms. The architecture becomes a form of social engineering, shaping not only how individuals move but how they perceive choice, freedom, and their relationship to technology. This process resonates with philosophical concerns about autonomy in mediated environments, where 'choice' is offered within constrained frameworks engineered for particular ends. The subtlety of these shifts makes them especially insidious, as users may conflate convenience with consent, mistaking algorithmic steering for natural preference.

Nonetheless, not all behavioural architecture is inherently coercive or negative. Thoughtful design can harness nudges to promote beneficial outcomes—such as encouraging eco-friendly routes, enhancing safety through real-time warnings, or supporting social inclusion by improving service in underserved areas. The challenge lies in ensuring transparency, user empowerment, and alignment with public interest rather than narrow commercial gain. Regulatory frameworks and ethical design standards have a crucial role in mediating these competing priorities, demanding that AV behavioural architectures be subject to rigorous oversight and inclusive deliberation.

Emerging research and pilot projects offer insights into how such balanced approaches might unfold. Participatory design processes that involve diverse stakeholders can surface hidden biases and broaden accountability. Algorithmic audits and open data initiatives contribute to demystifying behavioural nudges and fostering trust. Meanwhile, advances in explainable AI promise interfaces that can communicate decision rationales to users, supporting more informed and reflective engagement. These developments suggest that behavioural architecture need not be a one-way imposition but could evolve into a dialogic process enhancing autonomy even within technologically mediated mobility.

Ultimately, the behavioural architecture embedded in AV systems is a mirror reflecting broader societal tensions around technology, power, and freedom. It exposes the paradox of a mobility future that promises liberation through automation while simultaneously embedding new forms of control and influence. Engaging critically with this architecture is imperative—not only to safeguard individual autonomy but to ensure that the transformative potential of autonomous mobility serves a just and equitable vision of society. Every journey in an AV is therefore more than a transit from point A to B; it is a passage through a carefully constructed landscape of behavioural influence that demands our careful attention and ethical scrutiny.

The Risk of Over-Optimized Life

The pursuit of efficiency, long celebrated as a hallmark of modern progress, has acquired new urgency in the age of autonomous vehicles (AVs). These systems promise to streamline mobility, reducing travel times, fuel consumption, and human error through finely tuned algorithms and data-driven optimisation. Yet beneath the allure of seamless, frictionless movement lies a less obvious—and potentially profound—trade-off: the erosion of resilience and reflective capacity in human life. Efficiency-driven design, when taken to its logical extreme, risks cultivating a mode of existence that is overly optimised, brittle, and deficient in the very qualities that sustain adaptability and meaningful experience.

This tension is not merely technical but deeply philosophical. Efficiency suggests the elimination of redundancies, delays, and interruptions—factors traditionally viewed as obstacles to productivity. However, many of these 'frictions' are precisely what allow for spontaneity, error correction, and creative problem-solving. The reduction of variability in daily navigation and decision-making, a natural consequence of AV route optimisation and system automation, may inadvertently restrict the cognitive and emotional space necessary for reflection, serendipity, and resilience in the face of unforeseen circumstances.

Empirical studies in human factors and cognitive science reveal that reliance on optimised technological systems can dull situational awareness and reduce the capacity for improvisation. For instance, the automation of driving tasks can lead to what is known as "automation complacency," where operators become passive monitors rather than active participants, diminishing vigilance and delaying responses when intervention becomes necessary. AVs, by removing the human element from the control loop, extend this phenomenon beyond individual drivers to entire urban mobility networks. The resulting behavioural conditioning may foster environments where users expect perfect predictability and rapid problem resolution, inadvertently undermining their preparedness for disruption or error.

Moreover, the implications extend beyond cognition to the social and cultural realms. Travel behaviours shaped by algorithmic optimisation tend to homogenise routes and timing, marginalising local knowledge and diminishing opportunities for casual encounters or engagement with the environment. This standardisation risks eroding community fabric and the richness of urban life. The disappearance of 'shortcuts' and personalised detours, traditionally repositories of experiential knowledge, reflects a broader loss of agency that parallels concerns about the "death of expertise" in automated decision contexts. *The consequence is not only a loss of individual autonomy but a collective narrowing of the spatial and temporal imagination that underpins vibrant societies.*

The "over-optimised life" also raises questions about resilience in the face of systemic shocks. Efficiency often prioritises lean operations with minimal redundancy, leaving little margin for error. In transport systems, this might translate to reduced buffer times, tightly scheduled flows, and the elimination of alternative routes. While optimal under normal conditions, such configurations become fragile when confronted with unexpected disruptions—be it technical failures, natural disasters, or social unrest. Historical analyses of complex systems, from financial markets to infrastructure networks, show that over-optimisation can precipitate cascading failures, as tightly coupled elements lack the slack necessary to absorb shocks. The same principles apply to AV-enabled mobility, where the relentless pursuit of efficiency may paradoxically increase vulnerability.

These risks invite a reconsideration of what constitutes 'good design' in autonomous mobility. Rather than maximising efficiency alone, there is growing advocacy for incorporating principles of resilience, redundancy, and user agency. Some experimental frameworks suggest hybrid approaches that balance automated optimisation with opportunities for human intervention and discretionary choice. For example, AV interfaces might be designed to offer users meaningful control over route selection, even if these choices are less 'efficient' in conventional terms. Similarly, urban planning could integrate contingency spaces and temporal buffers to accommodate variability and promote social interaction.

Importantly, embracing imperfection and variability as design virtues aligns with longstanding philosophical insights about human flourishing. Thinkers such as John Dewey and Hannah Arendt emphasised the value of unpredictability and active engagement in cultivating reflective thought and meaningful action. The push toward frictionless, perfectly managed mobility risks suppressing these essential human qualities, substituting convenience for contemplation. In this light, the "death of the shortcut" becomes emblematic not only of changing travel patterns but of a deeper cultural shift—one that demands careful scrutiny and deliberate countermeasures.

At the policy level, addressing the risks of over-optimisation requires frameworks that prioritise transparency, inclusivity, and long-term societal wellbeing over narrow efficiency metrics. Regulatory agencies might enforce standards mandating resilience assessments or user autonomy provisions in AV system designs. Public discourse and participatory governance can ensure diverse voices shape the normative values embedded in these technologies. The integration of social science research and human-centred design methodologies stands as a crucial step in rebalancing technological potential with human dignity.

The dialogue on efficiency and its discontents is not merely academic; it has immediate practical stakes as AV deployment accelerates globally. Early adopters and policymakers face choices that will define mobility ecosystems for decades. The allure of flawless optimisation must be tempered by an appreciation of human complexity and the unpredictable nature of life on the move. By recognising and mitigating the risks inherent in over-optimised systems, society can foster autonomous mobility that supports not just speed and convenience but resilience, reflection, and the enduring human capacity to navigate uncertainty with creativity and care.

Exit Options and Backdoors

Autonomous vehicles (AVs) are heralded for their potential to revolutionise transport by offering seamless, efficient, and safe mobility. Yet beneath the promise of hands-free navigation and algorithmic precision lies a less examined and deeply consequential reality: the erosion of user autonomy through limited override capabilities and constrained exit options. The systems that govern AV behaviour are not merely passive tools but active agents of control, designed with layers of technical, legal, and commercial constraints that restrict human intervention. This phenomenon poses profound questions about freedom, responsibility, and the very nature of agency in an automated world.

At the heart of this issue is the tension between automation's efficiency and the human desire for control. Traditional vehicles afford drivers the ability to exert immediate, tactile influence over their motion—steering, accelerating, braking, or aborting journeys on a whim. Autonomous systems, by contrast, operate within tightly defined parameters and protocols that limit such intervention. For safety and reliability, manufacturers often design AVs with "fail-safe" modes that default to minimal or no user override during critical moments. While this architecture aims to mitigate accidents caused by human error, it simultaneously diminishes the driver's capacity to act according to judgment or preference.

This restriction is further compounded by the software-centric nature of AVs, which introduces a new dimension of "backdoors"—undocumented or non-transparent pathways by which manufacturers, regulators, or remote operators can intervene, update, or disable vehicle functions. These backdoors, while often justified for security patches or emergency control, raise significant concerns about surveillance, data privacy, and loss of user sovereignty. The prospect that an external agent might commandeer or restrict vehicle operation without the occupant's consent unsettles conventional notions of ownership and self-determination.

Real-world case studies illustrate these dynamics vividly. In trials where AVs have encountered ambiguous or unprecedented situations, remote operators have occasionally overridden onboard decisions to prevent accidents. Yet this layered control introduces latency, dependency, and legal ambiguity: Who bears responsibility when a remote command leads to harm? Furthermore, users accustomed to direct control may experience frustration, helplessness, or mistrust when the system refuses to relinquish authority, even when manual intervention seems desirable. This erosion of agency is not trivial; it affects how individuals relate to technology and their own autonomy.

The commercial incentives shaping AV design also influence exit options and override mechanisms. Manufacturers and service providers have vested interests in maintaining system integrity and limiting liability, which can translate into restrictive user agreements and technical barriers to modification or repair. The trend toward "software as a service" models compounds this dynamic, as vehicles become platforms subject to continual remote management rather than discrete consumer possessions. In this context, exit options are curtailed not only at the level of physical control but also through contractual and informational asymmetries.

Philosophically, these developments prompt reflection on the conditions for meaningful autonomy in technologically mediated environments. Autonomy entails the capacity for self-governance, encompassing both the freedom to act and the ability to reflect critically on those actions. When AV systems systematically circumscribe the latter—by limiting choices, obscuring decision logic, or precluding manual override—they risk reducing users to passive recipients of automated directives rather than active participants in their mobility. This shift resonates with broader societal debates about algorithmic governance and the "black-boxing" of decision-making.

There are, however, proposals and design philosophies that seek to recalibrate this balance. Human-centred design advocates argue for "negotiated autonomy" frameworks, where control is shared dynamically between user and machine, with transparent interfaces that inform and empower users to intervene when appropriate. These models emphasise the importance of preserving meaningful exit options, including clear manual override protocols, emergency disengagement, and user education. Regulatory frameworks are also evolving to require minimum standards for user control and system transparency, though enforcement remains uneven and contested.

The implications of limited exit options and backdoors extend beyond individual experience to encompass legal, ethical, and social dimensions. Legislators must grapple with liability attribution in hybrid control scenarios; ethicists must consider the legitimacy of surrendering agency to opaque systems; and societies must confront the potential for technological disenfranchisement, where certain populations are disproportionately affected by rigid or opaque AV controls. Ensuring equitable access to autonomy in mobility thus becomes a critical dimension of justice.

Ultimately, the challenge posed by exit options and backdoors in AV systems is emblematic of a larger paradigm shift in human-technology interaction. As automation permeates essential domains of life, the architecture of control and the preservation of autonomy demand urgent and sustained attention. Navigating this terrain requires a nuanced understanding that integrates technical constraints, user experience, and normative values. *Designing autonomous vehicles that respect and enable human agency—rather than supplant it—is not merely a technical challenge but a profound social imperative.*

Chapter 9
Global Divergences

Autonomous vehicles are more than feats of engineering or conveniences for urban commuters; they have emerged as instruments of geopolitical influence, shaping how nations exercise sovereignty, regulate authority, and project power on the international stage. The adoption and governance of AVs expose profound disparities among countries, reflecting a complex interplay of political philosophy, economic ambition, and strategic calculation. This chapter explores the variegated global landscape of AV integration, where divergent laws, cultural norms, and regulatory priorities produce a fragmented yet consequential technological ecosystem. To appreciate these differences is to recognise that technology, though borderless in potential, remains tightly interwoven with national identity and state control.

Regulatory approaches to AVs vary widely. In some liberal democracies, the emphasis is on safety, privacy, and public engagement, seeking to balance innovation with individual rights. Elsewhere, governments prioritise speed of deployment, economic growth, or enhanced surveillance capacities, sometimes at the expense of transparency or consent. These choices reveal competing visions of technological stewardship—whether the state functions as protector of citizen autonomy or as enforcer of centralised authority. The contrast between these pathways mirrors broader tensions between liberal and authoritarian governance, with each imprinting its values on the streets, both physical and digital, of the future.

Surveillance serves as a particularly revealing lens on AV governance. Authoritarian regimes often exploit autonomous vehicles not merely for transport efficiency but as extensions of a pervasive monitoring apparatus, integrating location tracking, facial recognition, and behavioural analytics into daily life with unprecedented precision. Democracies, by contrast, wrestle with public accountability and legal constraints, creating limits—albeit imperfect—on how AV-generated data can be used for control. These divergent applications underscore the delicate balance between security and liberty that each society negotiates in the age of automation.

Beyond domestic concerns, AVs are increasingly arenas of international competition. Vehicle software carries embedded cultural norms, safety assumptions, and behavioural priorities; when exported, these systems effectively disseminate ideology, shaping driving practices, regulatory expectations, and social norms across borders. The apparent neutrality of technology dissolves, revealing digital infrastructures as vectors of power, influence, and standard-setting in a globalised yet ideologically diverse landscape.

This dynamic sits squarely within the broader race for technological supremacy. AVs are not merely commercial assets; they have become symbols of national modernity and instruments in broader contests for AI, data infrastructure, and industrial dominance. Investments in AV research, deployment, and regulation are strategic moves, influencing supply chains, regulatory regimes, and international norms that will govern autonomous mobility for decades.

Real-world examples illuminate these contrasts vividly. China pursues a state-directed, rapid deployment model; the European Union adopts precautionary, highly regulated frameworks; and the United States navigates a patchwork of federal and state jurisdictions. Emerging economies face unique challenges, negotiating AV integration amid infrastructural constraints, governance complexities, and questions of technological dependency. Each approach reflects national priorities, and collectively, they underscore AVs' role as mirrors of political intent and as instruments of global negotiation.

Philosophically, this divergence raises pressing questions about standards, ethics, and sovereignty. Can universal norms coexist with culturally contingent values? Who decides the rules of the autonomous ecosystem, and whose interests are advanced or marginalised in that process? These are not abstract queries; they shape the lived experience of AV technology and the broader societal consequences of its deployment.

Through detailed analysis, this chapter equips readers to navigate the geopolitical currents shaping autonomous vehicles. It emphasises that the future of AVs is as much about social, political, and cultural negotiation as it is about engineering precision. Understanding these global divergences is essential for envisioning a future of autonomous mobility that is equitable, ethical, and resilient in the face of technological, ideological, and regulatory complexity.

Different Roads, Different Rules

The regulatory landscapes governing autonomous vehicles (AVs) reveal as much about a nation's political character and economic ambitions as they do about technological progress. Far from a uniform global phenomenon, the adoption and control of AV technology unfold along distinctly varied paths, shaped by unique legal frameworks, cultural attitudes, infrastructural realities, and strategic priorities. These divergent approaches underscore that the future of autonomous mobility is not a singular trajectory but a collection of competing experiments in governance, each with its own implications for safety, innovation, and public trust.

In the United States, regulatory fragmentation reflects the country's federal structure and its cultural valorisation of innovation and market freedom. The federal government delegates much authority to individual states, resulting in a patchwork of policies that range from permissive testing environments to more cautious, restrictive regimes. This decentralised model fosters rapid experimentation and private-sector leadership, yet it also generates inconsistencies that complicate interstate deployment and create regulatory uncertainty. Crucially, the U.S. approach often privileges commercial interests and technological advancement, sometimes at the expense of comprehensive safety oversight or data privacy protections. This has led to heated debates over liability, ethical programming, and the role of government in mitigating risks inherent in unproven technologies.

Contrast this with the European Union's approach, which exemplifies a more precautionary regulatory ethos. Here, AV deployment is tightly interwoven with broader commitments to privacy, environmental sustainability, and human rights. The EU's emphasis on transparency, public consultation, and harmonised standards aims to balance innovation with social responsibility. For instance, the General Data Protection Regulation (GDPR) imposes strict constraints on how AV companies collect and manage user data, a regulatory posture that significantly shapes AV development. Moreover, European regulators engage proactively with ethical frameworks, demanding that AV systems align with values such as fairness, accountability, and inclusivity. This comprehensive oversight reflects a political culture that views technology as a public good requiring robust democratic governance, rather than purely a market-driven commodity.

Meanwhile, China's approach illustrates a top-down, state-centric model prioritising rapid scale and integration of AV technology into national strategic objectives. The Chinese government's heavy investment in infrastructure, research, and pilot programmes is coupled with a permissive regulatory climate that accelerates deployment. However, this technological enthusiasm is inseparable from broader surveillance ambitions and social control mechanisms. AVs, in this context, become extensions of state power, integrated into extensive sensor networks and data ecosystems that enhance urban management and population monitoring. The government's willingness to subordinate individual privacy to collective order shapes not only how AVs operate but also public acceptance of pervasive data collection. This model offers an alternative vision of technological progress—one that foregrounds sovereignty and social stability over personal autonomy.

Emerging economies present yet another dimension of regulatory complexity. Countries in Latin America, Africa, and Southeast Asia confront infrastructural challenges and resource constraints that limit their capacity to regulate or deploy AVs comprehensively. Often, these regions become arenas for technology transfer, where imported AV systems encounter legal and social environments unprepared for their demands. This dynamic raises concerns about technological dependency and the imposition of foreign regulatory norms unsuited to local conditions. Furthermore, limited public engagement and weak institutional oversight risk exacerbating inequalities, as AV technologies primarily serve affluent urban populations while neglecting broader mobility needs. The regulatory void in such contexts underscores how AV governance intertwines with questions of development, equity, and global technological justice.

Beyond individual national policies, the international governance of AVs remains an evolving and contested domain. Organisations such as the United Nations Economic Commission for Europe (UNECE) have begun formulating guidelines and safety standards, yet the voluntary nature of these frameworks means their influence varies widely. The interplay between national sovereignty and supranational coordination introduces tensions about who sets the rules for technologies that inherently transcend borders. Meanwhile, private corporations, through their control of proprietary algorithms and data, exert disproportionate influence over the regulatory agenda, often outpacing government capacity to respond. This corporate–state dynamic complicates efforts to establish harmonised, equitable standards and highlights the political economy underlying autonomous mobility.

These varied regulatory experiments reveal conceptual tensions between innovation and control, market dynamism and social protection, national sovereignty and global standardisation. They compel reflection on how different societies value risk, autonomy, and technological progress—and how these values materialise in law and practice. *Regulatory divergence is not merely a technical matter but a reflection of profound normative choices about the kind of futures societies want to build and whom they wish to empower or exclude in the process.*

Understanding these differences is essential not only for anticipating where and how AVs will become integrated into everyday life but also for critically assessing the broader consequences of technology governance. The patchwork of "different roads, different rules" shapes not only the contours of autonomous mobility but also the politics of trust, accountability, and justice in a rapidly automated world. As AVs reshape urban landscapes and social relations, the regulatory frameworks underpinning them will determine whether these transformations enhance collective wellbeing or deepen existing fractures.

In tracing the diverse national pathways of AV regulation and adoption, this subchapter offers readers a nuanced comprehension of the geopolitical and cultural forces driving autonomous vehicle governance. It illuminates how law and policy serve as crucibles where technology meets society, exposing the variegated realities of a world negotiating the promises and perils of autonomy. *The contours of these differing roads reveal more than regulatory divergence—they map competing visions of progress, power, and possibility in the twenty-first century.*

Surveillance or Sovereignty?

Autonomous vehicles (AVs) are emerging as much more than a technological innovation in mobility; they have become instruments through which states project power, enforce control, and negotiate their visions of governance. The deployment of AV technology unfolds against starkly contrasting political backdrops, with democratic and authoritarian regimes harnessing its capabilities in profoundly different ways. These divergent uses expose fundamental tensions between the ideals of individual privacy and collective security, between citizen autonomy and state sovereignty, and between technological progress and political control. Understanding how AVs are integrated into these distinct frameworks illuminates the broader geopolitical stakes embedded within autonomous mobility's rise.

In authoritarian contexts, AVs are frequently entwined with comprehensive surveillance infrastructures that extend the state's capacity to monitor and manage populations. The deployment of sensor-laden vehicles, combined with pervasive urban monitoring networks, feeds real-time data into centralised databases that map citizen movements with unprecedented granularity. This data not only facilitates efficient traffic management but also functions as a tool of social control. The technological affordances of AVs—continuous location tracking, behavioural pattern analysis, facial recognition integration—amplify existing mechanisms of state power, enabling rapid identification and targeting of dissent. Governments may leverage AV-generated metadata to enforce curfews, restrict mobility in politically sensitive zones, or surveil minority communities under the guise of public safety. Such applications blur the line between transportation infrastructure and surveillance apparatus, embedding AVs within broader authoritarian strategies of population governance.

China's AV ecosystem epitomises this fusion of surveillance and sovereignty. The state's ambitious "smart city" projects incorporate autonomous vehicles as nodes in a vast, interconnected surveillance web, coalescing AV data with other intelligence sources such as social credit systems and biometric databases. The rapid scaling of AV deployment occurs alongside expansive legal frameworks granting the state sweeping authority to collect and utilise data, often with limited transparency or public recourse. This integration serves multiple purposes: enhancing urban efficiency, underpinning social stability, and reinforcing regime legitimacy through a technologically mediated paternalism. The state's control over AV infrastructure transforms mobility into an instrument of governance that simultaneously promises convenience and enforces compliance.

By contrast, democratic regimes confront a more complex balancing act, negotiating the integration of AVs within pluralistic societies that prize individual rights and legal protections. While the technological capabilities of AVs can likewise facilitate surveillance, democratic institutions impose legal and normative constraints designed to safeguard privacy, ensure accountability, and prevent abuses of power. Public debates often centre on consent, data protection, and the scope of government access to mobility data. Democratic governments typically adopt a cautious, deliberative regulatory approach that involves multiple stakeholders, including civil society actors, industry representatives, and privacy advocates. The challenge lies in harnessing AV technology's benefits—improved safety, reduced congestion, increased accessibility—without sacrificing the liberties and democratic oversight mechanisms that underpin societal trust.

Yet, the democratic use of AVs for power projection is not absent. Police departments in several Western countries have begun integrating AV-generated data streams into predictive policing and traffic enforcement operations. Insurance companies deploy driving behaviour analytics derived from AV sensors to calibrate premiums and detect fraud. While ostensibly benign, these applications raise profound questions about fairness, discrimination, and the disproportionate targeting of marginalised communities. The "smart" functionalities embedded in AVs—route optimisation, speed regulation, passenger monitoring—can inadvertently institutionalise biases encoded in algorithms or data sets. Democratic societies thus face a delicate dilemma: how to regulate and audit these technologies rigorously enough to prevent creeping surveillance or discriminatory practices while fostering innovation and public utility.

The international dimension further complicates the interplay between surveillance and sovereignty. Authoritarian and democratic powers alike are engaged in a global contest over the standards, norms, and governance regimes that will shape AV deployment worldwide. Exporting AV technologies becomes a vector for extending political influence, where embedded value systems and governance models accompany hardware and software. Authoritarian regimes may offer AV systems as part of broader "smart city" packages that enhance surveillance capabilities for client states, while democratic powers advocate for privacy-respecting frameworks aligned with human rights commitments. This competition underscores that AV technology is not merely a tool of mobility but a geopolitical instrument through which regimes export visions of governance, contest digital sovereignty, and recalibrate global power relations.

At its core, the use of AVs by differing political systems surfaces fundamental questions about the nature of sovereignty in the digital age. Autonomous vehicles generate detailed, dynamic traces of human movement—data that is both intensely personal and strategically valuable. How states claim authority over this data, regulate its use, and negotiate the boundaries of individual freedom reveals much about their political priorities. In authoritarian settings, the confluence of surveillance and mobility facilitates a mode of governance that prioritises order and control, often at the expense of privacy and civil liberties. In democracies, the challenge is to harness AV innovation while ensuring transparency, accountability, and respect for individual rights. Yet even these regimes risk expanding state power under the mantle of public safety and technological progress.

The tension between surveillance and sovereignty embedded in AV deployment is thus not simply a question of technology but a reflection of competing political imaginaries about authority, autonomy, and the social contract. As autonomous vehicles become integrated into the fabric of daily life, they crystallise these tensions in tangible, impactful ways. Whether

AVs evolve as tools of empowerment or instruments of control depends not only on technological design but on the political choices made by governments, regulators, and societies. This subchapter invites a critical examination of these choices, urging readers to consider how the governance of autonomous mobility intertwines with broader struggles over freedom, power, and the future of democratic life.

Exporting Standards, Importing Power

The rise of autonomous vehicles (AVs) is not merely a story of technological innovation; it is equally a tale of geopolitical influence and cultural projection. As AV companies expand their reach beyond domestic markets, they carry with them embedded value systems—normative choices encoded into algorithms, software architectures, and hardware designs—that inevitably shape the contours of global mobility norms. The seemingly neutral code that governs how an AV perceives, decides, and acts is in fact a carrier of cultural assumptions, regulatory philosophies, and political priorities. These embedded values, exported through commercial and strategic partnerships, assert a form of technological sovereignty that wields profound implications for international governance, power dynamics, and local autonomy.

Understanding how AV companies export standards is critical to grasping the broader politics of technology in the twenty-first century. Standards are not merely technical specifications; they constitute the invisible infrastructure of modern life, defining what is possible, permissible, and preferable within a given technological ecosystem. When multinational corporations develop AV systems tailored primarily to their home markets' legal frameworks, ethical expectations, and infrastructural realities, they create de facto templates that influence other nations' regulatory environments and user experiences. This process can promote harmonisation but also risks eroding local customs, regulatory discretion, and social norms under the guise of global efficiency and interoperability.

The path-dependency inherent in software development exacerbates this dynamic. AV technologies require extensive testing, data accumulation, and iterative refinement, often in collaboration with local governments and infrastructure providers. This generates a strong incentive for companies to replicate successful models across borders rather than redesign systems from scratch for each unique jurisdiction. The result is a standardisation of practices, data formats, sensor configurations, and decision-making logics that privilege certain values—such as risk tolerance, privacy expectations, and traffic rule enforcement—over others. For example, an AV calibrated for the United States' traffic laws and liability norms may struggle to adapt to the more stringent data privacy requirements or different road-user behaviours of the European Union or India. As these companies assert influence through market dominance or governmental partnerships, they effectively export the regulatory and ethical frameworks embedded in their products.

This exportation is not merely a market-driven phenomenon but also a deliberate exercise of soft power. States with dominant AV industries leverage these technologies to set international standards and norms, akin to the historic role of broadcast standards, telecommunications protocols, or internet governance models. The United States, with its strong Silicon Valley presence, exerts considerable influence in shaping global expectations around data use, liability regimes, and interoperability standards. China, pursuing its "New Infrastructure" and "Digital Silk Road" initiatives, actively promotes AV standards that align with its own surveillance-oriented governance and infrastructural priorities, exporting these through bilateral agreements and development projects. The convergence of corporate interests and state strategies creates a geopolitical contest where AV standards become instruments of influence, shaping not only technological landscapes but also political alignments.

Crucially, the values embedded in AV code extend beyond technical parameters to ethical and social dimensions. Decisions about how an AV negotiates ethical dilemmas—such as collision avoidance prioritisation, passenger versus pedestrian safety, or privacy trade-offs—reflect culturally specific moral philosophies. These normative decisions, encoded as algorithms, become universalised when exported globally, effectively imposing one society's ethical calculus on diverse populations with different moral frameworks. For instance, AVs developed in societies prioritising individual freedom and privacy may be less acceptable in communities that favour collective security or social harmony. Conversely, systems designed with heavy data surveillance may conflict with jurisdictions that have strong data protection laws and public expectations. Thus, the export of AV technologies carries implicit ethical colonisation, constraining local societies' ability to define their own mobility values.

Case studies illustrate these complexities. Waymo's gradual internationalisation reflects an effort to maintain American norms of user consent, privacy safeguards, and transparent liability. Meanwhile, Baidu's Apollo project exemplifies a different approach, integrating AV deployment with China's state-directed urban surveillance and data-sharing policies, exported through partnerships in Belt and Road countries. European AV initiatives often emphasise robust privacy frameworks and multi-stakeholder governance models, yet face challenges in influencing global norms due to market scale limitations. The uneven distribution of technological capacities, regulatory frameworks, and geopolitical clout results in a patchwork of AV governance that reflects underlying global power asymmetries.

This patchwork raises critical questions about sovereignty and technological autonomy. When nations adopt AV technologies embedded with foreign values, they risk ceding control over their regulatory sovereignty and cultural self-determination. The challenge is not merely technical adaptation but the political and ethical negotiations surrounding whose values are prioritised, who governs the data flows, and how local populations engage with these new forms of mobility. The AV standardisation process can either support a pluralistic approach that respects local contexts or enforce homogenised regimes that reflect dominant corporate-state coalitions. The stakes extend beyond roads and vehicles to the very architecture of social order in a digital age.

In sum, the exportation of AV standards is a potent form of power projection that transcends traditional geopolitical boundaries. *Technological code acts as a vector for cultural and political influence, shaping global norms in ways that redefine sovereignty and governance.* Recognising this dynamic compels policymakers, technologists, and civil society actors to engage critically with the values encoded in AV systems and advocate for regulatory frameworks that preserve local autonomy and democratic accountability. As autonomous vehicles become fixtures of global transport networks, the ethical, political, and cultural implications of their embedded standards demand deliberate attention, lest the future of mobility become a conduit for unequal power relations disguised as technological progress.

The Race for the Autonomous Future

Autonomous vehicles (AVs) are no longer merely a technological marvel or a symbol of futuristic convenience—they have become a decisive arena in the geopolitical struggle for technological supremacy. This contest transcends industry competition, embedding itself deeply in national security concerns, economic strategy, and global influence. Governments and corporations alike recognise that leadership in AV technology confers advantages far beyond transportation efficiency: it is a powerful instrument of strategic leverage, shaping alliances, trade flows, and the very architecture of urban life on a planetary scale. The race to dominate the autonomous future thus resembles an arms race, where technological innovation is weaponised as a means to assert power and secure competitive advantage.

The stakes are exceptionally high because AVs sit at the intersection of multiple critical domains: artificial intelligence, big data analytics, cybersecurity, and advanced manufacturing. Mastery of AV technology signals not only industrial prowess but also the ability to control data ecosystems and set standards that influence global regulatory frameworks. This is a race whose winners will dictate how cities operate, how citizens move, and how information flows across borders. Consequently, AVs are rapidly becoming a geopolitical tool as significant as traditional military assets, casting a shadow far beyond the factory floor or test track.

One cannot overstate the strategic imperative behind national AV initiatives. The United States, China, and the European Union have each articulated comprehensive policies linking autonomous technology development to broader goals of economic resilience, military capability, and digital sovereignty. China's "New Infrastructure" strategy, for instance, prioritises AV deployment within its broader digital ecosystem ambitions, integrating smart transportation with surveillance networks and data governance models that reinforce state control. The US approach blends private-sector dynamism with military interest, as demonstrated by the Pentagon's investments in AVs for logistics and combat support. The European Union, in contrast, emphasises regulatory coherence, ethical AI, and interoperability, seeking to balance innovation with citizen rights and sustainability. These divergent strategies reflect underlying political philosophies and power structures, yet all share a recognition that AV leadership is vital for future influence.

Beyond individual national programmes, the international landscape is marked by contestation over standards, intellectual property, and market access. AV technologies rely on vast, high-precision data sets and sophisticated sensor networks, resources often bound by national regulations and geopolitical tensions. Control over these inputs and the ensuing algorithms becomes a source of leverage. For example, the competition to develop next-generation 5G and 6G networks is inextricably linked to AV performance, with dominant players seeking to lock competitors out through infrastructure and spectrum control. Similarly, the ownership of mapping data and real-time traffic information confers strategic advantage, as these resources underpin AV navigation and safety. The race thus encompasses not just vehicles themselves but the entire digital and physical infrastructure that sustains them.

The arms-race analogy extends to the potential militarisation and dual-use nature of AV technologies. Autonomous navigation and decision-making algorithms developed for civilian transport have clear applications in defence contexts, from unmanned ground vehicles to logistical automation. This dual-use characteristic intensifies the competition, blurring the lines between commercial innovation and national security. The rapid pace of AV advancement raises difficult questions about regulatory oversight, export controls, and the risks of proliferation. As nations invest heavily to outpace rivals, the prospect of an unstable technological escalation looms—one where deployment speed may override considerations of safety, ethics, or global cooperation.

Nevertheless, the race for AV dominance is not solely a zero-sum contest. Collaborative initiatives also exist, driven by shared interests in safety standards, environmental benefits, and equitable access. Multilateral bodies, such as the United Nations Economic Commission for Europe and the International Organization for Standardization, facilitate dialogue on harmonising regulations and ethical guidelines. However, these efforts must navigate competing national priorities and corporate interests that often impede consensus. The balance between competition and cooperation will shape the trajectory of AV technology development and its integration into society.

Crucially, the geopolitical contest over AVs cannot be disentangled from broader technological and economic transformations. AV leadership is both a reflection and a driver of shifts in global supply chains, labour markets, and urban planning. Control over the autonomous future thus confers influence over the 21st-century economic order, redefining centres of wealth and innovation. The interplay of state ambition, corporate strategy, and societal impact forms a complex mosaic, where technology, power, and identity intersect.

The race for the autonomous future is a defining strategic challenge of our time. It demands that policymakers, technologists, and civil society confront not only questions of engineering and economics but also those of governance, ethics, and international relations. *Mastering AV technology will determine who sets the rules of tomorrow's mobility, who controls critical infrastructure, and who wields influence in a world increasingly defined by data and automation.* As this contest unfolds, its outcomes will reverberate across societies, economies, and geopolitical orders, underscoring the profound entanglement of technology and power in the modern age.

Chapter 10
Reclaiming the Human Lane

Rethinking movement as a fundamental right requires a profound reimagining of how societies organise and govern mobility. For too long, transportation has been treated as a privilege—contingent on market forces, patchy infrastructure, and regulatory frameworks that often leave the most vulnerable at a disadvantage. Yet the rise of autonomous vehicles is transforming not just the mechanics of travel, but the very structures through which access is determined. In this context, the argument for recognising safe, reliable, and dignified mobility as a public right becomes pressing. Equitable access is no longer a technical question alone; it is a matter of justice, inclusion, and the conditions that allow human flourishing. Policymakers, urban planners, and technologists alike are called upon to design systems that extend the benefits of AVs beyond narrow economic interests to serve the broader community.

Emerging from this conversation is the concept of humane autonomy—a design philosophy that resists the wholesale surrender of control to algorithms. Humane autonomy insists that technology must align with human values, embedding transparency, ethical responsiveness, and avenues for meaningful user intervention. In doing so, it reasserts the role of the individual as an active participant rather than a passive recipient, balancing the efficiencies of automation with the unpredictability and nuance of everyday life. This approach challenges the prevailing narrative of technological inevitability, advocating for systems that respect the rhythms and complexity of human existence rather than imposing rigid, efficiency-driven templates.

Central to this dialogue is the question: what should remain beyond the reach of automation? Some domains of human experience—empathy, creativity, moral judgment—resist mechanisation without a significant loss of meaning. By defining these protected spaces, societies can safeguard a "human lane," preserving the zones where interpersonal connection, ethical deliberation, and authentic agency endure. This is not a rejection of technological progress but a deliberate assertion of autonomy over the conditions under which it operates, ensuring that efficiency does not come at the cost of humanity.

Living alongside the silent chauffeur—the omnipresent yet invisible AV—demands the cultivation of new mental models and civic strategies. Individuals must learn to navigate the subtleties of delegation without surrendering vigilance, while communities must establish frameworks for accountability, public deliberation, and inclusive governance. These literacies and strategies empower people to engage critically with automation, transforming what could be a passive experience into one guided by awareness and agency. The goal is a form of cohabitation with technology that preserves human values while harnessing the efficiencies AVs can offer.

This chapter explores these intertwined concerns in depth, framing mobility not simply as movement from one point to another, but as a profound social and ethical right. It charts pathways toward reclaiming the human lane—spaces where technology serves, rather than constrains, humanity—and encourages reflection on how societies might shape autonomous futures that uphold freedom, dignity, and collective well-being. The discussion presented here aims to provoke critical thought, deepen understanding, and inspire actionable frameworks for those tasked with creating a just and humane mobility landscape for generations yet to come.

Rethinking Movement as a Right

The premise that movement is a fundamental right challenges deeply ingrained perceptions of transportation as a mere commodity or convenience. Across many societies, the ability to travel—whether for work, education, healthcare, or social connection—remains unevenly distributed, often determined by geography, socioeconomic status, or infrastructural priorities. Autonomous vehicles (AVs), while heralded as transformative agents of mobility, risk exacerbating existing inequalities unless their integration is guided by a robust commitment to equitable access. To reimagine movement as a right is to insist that mobility infrastructure and policy must serve all members of society, embedding transportation within the broader matrix of social justice and public welfare rather than relegating it to market dynamics alone.

Historically, the discourse on transportation rights has been overshadowed by the privileging of private car ownership and profit-driven service models. This paradigm has produced sprawling urban forms, underfunded public transit, and exclusionary zoning practices that marginalise vulnerable populations. Yet, mobility underpins the exercise of numerous other rights—access to employment, participation in civic life, and even basic survival. By reframing movement as a public right, policymakers confront the ethical imperative to dismantle barriers and construct inclusive systems that acknowledge transportation as a social good. This shift requires not only infrastructural investment but also the recalibration of regulatory frameworks to prioritise universal access and affordability.

Contemporary policy initiatives offer illustrative examples of this evolving mindset. In countries such as Finland and Canada, transportation equity has been enshrined in strategic national plans, linking mobility to health outcomes and economic participation. Localised programs, including subsidised public transit and community ride-sharing schemes, demonstrate practical pathways toward reducing mobility deserts. These examples, however, often remain fragmented, highlighting the necessity of comprehensive policy frameworks that harmonise federal, regional, and municipal efforts. Autonomous vehicles, with their promise of enhanced accessibility, could catalyse this transformation—but only if governed by principles that ensure inclusivity rather than perpetuate exclusion.

Central to this discussion is the recognition that the deployment of AVs must be accompanied by stringent safeguards against deepening digital and infrastructural divides. The risk that AV technology serves predominantly affluent, urban populations—while neglecting rural or low-income communities—is significant. Moreover, the economics of AV operation may prioritise profitability over service to the underserved, reinforcing patterns of mobility injustice. To counter this, regulatory mechanisms such as universal service obligations, equitable fare structures, and mandates for inclusive design are critical. These policies must be crafted with sensitivity to local contexts, acknowledging that mobility rights manifest differently across diverse demographic and geographic landscapes.

Philosophically, asserting movement as a right resonates with broader conceptions of freedom and citizenship. Mobility enables agency—the capacity to navigate one's environment, make choices, and engage fully in society. Denying or restricting movement undermines these foundational freedoms, engendering forms of social exclusion that manifest in economic precarity, health disparities, and cultural isolation. Thus, transportation policy intersects profoundly with human rights frameworks, demanding that the governance of emerging technologies like AVs be subjected to rigorous ethical scrutiny. This perspective challenges technocratic narratives that emphasise efficiency or innovation at the expense of social equity.

The integration of AVs also prompts critical inquiry into the role of the state versus the market in ensuring mobility rights. While private sector innovation drives much of the technological advance, reliance on market mechanisms alone risks commodifying movement in ways that exclude those unable to pay. Public sector stewardship is essential not only for regulation but also for direct provision of services and infrastructure. The tension between fostering innovation and safeguarding public interest necessitates nuanced governance models that leverage the strengths of both sectors while holding private actors accountable to social objectives.

In practice, this rethinking calls for novel policy instruments and collaborative governance structures. Examples include mobility-as-a-service (MaaS) platforms that integrate diverse transport modes under unified access schemes, designed to prioritise equity and user empowerment. Legislative frameworks might mandate transparency in AV algorithms to prevent discriminatory routing or service allocation. Investment strategies could direct subsidies toward underserved regions, ensuring that the autonomous future is not confined to elite enclaves. Importantly, these efforts require ongoing public engagement and participatory policymaking to reflect the lived realities of communities most affected by mobility inequities.

Understanding movement as a public right also invites attention to environmental and urban sustainability. Equitable mobility systems can mitigate the negative externalities of car-centric urban sprawl, promoting walkability, shared transportation, and reduced emissions. Autonomous vehicles, integrated thoughtfully, have the potential to support these goals by optimising routes and encouraging shared rides—if underpinned by policies that align technological capabilities with social and environmental imperatives. The framing of movement as a right thus aligns with wider societal commitments to sustainable development and climate justice.

In sum, reimagining transportation as a public right situates mobility at the heart of democratic governance and social equity. Autonomous vehicles, while offering unprecedented technological possibilities, must be harnessed within frameworks that prioritise universal access, dignity, and inclusion. This entails a deliberate policy architecture that balances innovation with ethical responsibility, ensuring that the promise of AVs is realised not only for some but for all. Only by anchoring movement in the rights discourse can societies avoid replicating past inequalities and instead foster genuinely accessible, just, and humane mobility futures.

Building Humane Autonomy

Autonomous vehicles (AVs) promise to revolutionise transportation, offering unprecedented efficiency, safety, and accessibility. Yet beneath this technological optimism lies a crucial question: how can these systems be designed to preserve human dignity and agency? The challenge extends far beyond engineering safe navigation; it demands a re-examination of the relationship between humans and machines, ensuring that autonomy does not devolve into abdication of control. *Building humane autonomy* means crafting AVs that respect user autonomy, foster trust without fostering dependency, and embed ethical principles that protect the diverse ways in which people engage with mobility.

The concept of autonomy, particularly in transportation, is inherently linked to freedom and responsibility. Driving is not merely a mechanical task but a deeply embodied, cognitive, and social experience. To cede control to an algorithm is to alter this relationship fundamentally. While AVs can reduce human error and increase accessibility for those unable to drive, they also risk diminishing the individual's active participation in movement, which is central to self-efficacy and personal dignity. Designers and policymakers must therefore confront the tension between machine autonomy and human control, resisting narratives that valorise seamless automation at the expense of user empowerment.

One core principle in building humane autonomy is *transparent agency*. Users must understand not only that the vehicle is operating autonomously but also the logic and limitations underlying its decisions. This transparency requires interfaces that communicate real-time information clearly and intuitively, enabling users to anticipate and intervene if necessary. Consider the experimental user interfaces in the AV sector that display sensor data and route rationale on dashboards. While some AVs currently offer minimal feedback, future designs should prioritise interactive transparency to sustain a user's situational awareness. Without such mechanisms, autonomy becomes opaque, inviting disengagement or dangerous overreliance.

Equally important is the *preservation of meaningful choice*. Rather than imposing rigid control algorithms, AV systems should allow for user preferences and overrides that acknowledge varied risk tolerances, cultural contexts, and individual habits. For example, some users might prioritise the fastest route, others the most scenic, or the one with the fewest intersections. Humane autonomy respects these preferences and accommodates human judgment rather than supplanting it entirely. This principle challenges the prevailing design ethos that often treats optimisation solely in terms of efficiency or cost reduction, ignoring the subjective and qualitative dimensions of travel.

The integration of *ethical reflexivity* into AV design further exemplifies humane autonomy. Autonomous systems inevitably encode values—how to prioritise pedestrian safety, react in emergency situations, or weigh the rights of different road users. These decisions are not merely technical but profoundly moral. Designing AVs that can reflect on these ethical trade-offs, adapt to social norms, and remain accountable requires cross-disciplinary collaboration among engineers, ethicists, sociologists, and end-users. It also necessitates regulatory frameworks that mandate ethical transparency and public engagement, avoiding opaque "black box" decision-making that alienates users.

A case in point is the debate over the "trolley problem" in AV ethics. While often criticised as a philosophical thought experiment of limited real-world applicability, it underscores the importance of deliberate ethical programming and user involvement. Should an AV prioritise the safety of its passengers over pedestrians? How much agency should users have in setting these parameters? Designing humane autonomy entails not only programming such decisions but also communicating them openly and incorporating user feedback, thus preserving trust and dignity.

Beyond interface and ethics, *human-centred adaptability* is essential. AVs must accommodate diverse users, including those with disabilities, the elderly, and those unfamiliar with digital technologies. For instance, vehicle controls and feedback systems should be adjustable to varying sensory and cognitive abilities, ensuring inclusivity. Furthermore, adaptability extends to social contexts: autonomous vehicles operate within complex urban environments shaped by cultural behaviours, legal frameworks, and infrastructural constraints. Systems that can learn from and respond to local customs—such as yielding norms or pedestrian behaviours—enhance both safety and social acceptability.

Moreover, *maintaining human oversight* is pivotal in avoiding the pitfalls of automation complacency. Research in human factors psychology reveals that passive supervision often leads to degraded vigilance and slower reaction times, a phenomenon termed "automation complacency." Humane autonomy addresses this by designing shared-control models where human drivers and AV systems collaborate dynamically, rather than the latter assuming complete control. Such models might involve graduated automation levels, where human drivers remain engaged through tactile feedback or periodic control requests, preserving cognitive skills and situational awareness.

Several industry initiatives point toward this shared autonomy model. For example, Tesla's Autopilot and General Motors' Super Cruise offer semi-autonomous driving where human drivers retain ultimate responsibility. However, these systems also highlight risks of overtrust and disengagement, suggesting that design alone is insufficient; regulatory oversight and user education must accompany technological development. Indeed, humane autonomy calls for a holistic approach integrating technology, policy, and human factors science.

Lastly, *designing for dignity* means respecting the social and emotional dimensions of transportation. Riding in an AV is not a neutral experience—it involves vulnerability, anticipation, and the need for control in unfamiliar or stressful situations. Vehicles that provide intuitive control options, clear communication, and respectful interfaces contribute to a sense of security and self-respect. This extends to the spatial design of interiors, offering privacy and comfort rather than cold, clinical environments. The psychological impact of relinquishing control is profound, and humane autonomy acknowledges this by centring the human experience.

In practice, achieving these principles requires systemic change across the AV ecosystem. Manufacturers must embed ethical design early in development cycles, regulators must enforce standards that mandate transparency and user control, and communities must be engaged in co-creating mobility futures. Success hinges on resisting reductive conceptions of autonomy as mere automation, and instead embracing a nuanced, human-centred vision where technology amplifies rather than replaces human agency.

Ultimately, *building humane autonomy* is an ethical and design imperative that shapes not only how autonomous vehicles function but how societies envision the future of mobility. It demands balancing technological innovation with human dignity, safeguarding agency while harnessing efficiency. As autonomous vehicles move from prototypes to everyday reality, their capacity to respect and empower users will determine whether they fulfill their transformative promise or perpetuate alienation in motion.

What Should Never Be Automated?

The relentless advance of automation has prompted a profound reckoning: not every facet of human life should, or indeed can, be surrendered to machines. As autonomous systems infiltrate diverse domains—from manufacturing floors to financial markets, from healthcare diagnostics to personal transportation—the question arises with renewed urgency: what must remain resolutely human? This inquiry transcends mere technical feasibility or economic efficiency; it probes the very foundations of identity, ethics, and social cohesion. To decide what should never be automated demands a sober, nuanced reflection on the limits of algorithmic governance and the irreplaceable qualities of human judgment, empathy, and moral responsibility.

Automation, when wielded with precision and care, yields remarkable benefits—reducing drudgery, enhancing safety, and expanding access. Yet it carries inherent risks of dehumanisation, loss of agency, and erosion of skills. The initial allure of handing over complex tasks to artificial intelligence often obscures subtle but consequential trade-offs. For instance, a machine may flawlessly execute predefined procedures but fail to grasp the context, nuance, or moral subtleties that a human actor perceives intuitively. In domains laden with ethical ambiguity and interpersonal dynamics, automation can strip away vital elements of human engagement, leaving cold, brittle systems prone to unintended harm.

One fundamental criterion for resisting automation is the *preservation of human dignity*. Tasks that intimately involve care, trust, and relational judgement—such as parenting, counselling, or nursing—are anchored in empathy and emotional intelligence that no algorithm can authentically replicate. Consider the controversy surrounding robot caregivers for the elderly: while they may assist with physical tasks, their inability to offer genuine companionship or respond to emotional needs risks compounding loneliness and alienation. Similarly, legal decision-making, despite the appeal of data-driven efficiency, demands human discernment that weighs justice beyond mere rules or statistics. The subtlety of interpreting intent, mitigating bias, or recognising individual circumstances defies simplistic automation.

Closely related is the domain of *moral accountability*. Automated systems lack consciousness and ethical agency; when errors occur, responsibility becomes diffuse. Delegating decisions with profound consequences—whether in medical triage, autonomous vehicle ethics, or military applications—requires human oversight to uphold accountability. The "black box" nature of many machine-learning models complicates this, as opaque algorithms produce outcomes without transparent rationales. Entrusting machines with such power risks absolving human agents of their duties, undermining social trust and potentially enabling abuses. The persistence of human control, even if mediated through technology, safeguards against such erosion.

Another area warranting exemption from automation is *creative and existential expression.* Art, literature, and other forms of cultural production arise from uniquely human experiences—imagination, suffering, joy, and the search for meaning. While AI can mimic stylistic patterns or generate novel combinations, it cannot embody the lived reality that imbues creative acts with significance. The mechanisation of creativity risks commodifying culture, diluting authenticity, and constraining diversity. In this vein, the human capacity for reflection and moral imagination is essential for navigating a world where automated systems increasingly shape environments and choices.

The resistance to automation also stems from preserving *skills critical to human autonomy.* Overdependence on machines can atrophy capacities such as problem-solving, situational awareness, and critical thinking. In transportation, for instance, fully autonomous vehicles may erode drivers' abilities to react in emergencies, compounding risks when automation fails. Analogously, reliance on automated financial advisors might dull individual understanding of personal economics, reducing resilience in volatile markets. Retaining human involvement in essential skills not only enhances safety but sustains agency and competence.

Social and political dimensions further complicate automation choices. Technologies are embedded within power structures and cultural contexts that shape their impact. Automating public services, policing, or welfare without inclusive dialogue risks entrenching inequalities and marginalising vulnerable groups. Decisions about what to automate must be democratised, reflecting societal values rather than narrow commercial interests or technocratic imperatives. This calls for participatory governance models and ethical frameworks that protect collective rights and foster equitable outcomes.

It is crucial to recognise, however, that the boundary between what should and should not be automated is neither fixed nor universal. Different cultures, professions, and individuals hold varying conceptions of trust, responsibility, and the meaning of human work. Automation's encroachment may, in some cases, liberate individuals from hazardous or monotonous labour, enabling flourishing in other spheres. The challenge lies in discerning where technology complements human capacities and where it supplants essential human qualities. Such discernment demands ongoing, critical engagement across disciplines and communities.

Emerging frameworks in technology ethics, such as "human-in-the-loop" and "meaningful human control," seek to mediate this balance. These approaches emphasise hybrid systems where automation enhances rather than replaces human decision-making, with explicit channels for intervention and override. For example, in healthcare, AI-assisted diagnostics can support physicians but not supplant their final judgement and patient relationship. In autonomous vehicles, retaining driver engagement in critical situations preserves accountability and situational awareness. Such models respect both the power and the limits of automation.

Moreover, educational imperatives accompany these reflections. Societies must cultivate digital literacy and ethical reasoning to equip individuals to navigate increasingly automated environments. Empowering people to understand algorithmic processes, question their biases, and assert control is foundational to resisting passive surrender to automation. In doing so, education becomes a bulwark for preserving the human domains that machines should not claim.

Ultimately, resisting automation in certain domains is a matter of *protecting the essence of human experience.* It is an assertion that not all efficiency gains justify surrendering elements of moral judgment, emotional connection, creativity, and personal autonomy. The decisions we make today about automation will shape social relations, individual identities, and collective futures. Thus, a deliberate, reflective stance on what should never be automated is not technophobic resistance but an essential act of stewardship over the human condition in an increasingly automated world.

Living with the Silent Chauffeur

The advent of autonomous vehicles (AVs) signals a profound transformation in mobility, one that transcends mere technological innovation to touch on the core of human experience—how we move, interact, and exert agency in the spaces between origin and destination. Yet, the silent chauffeur that navigates without human hands or eyes presents challenges that extend beyond mechanics or safety. It requires us to develop new mental models and civic strategies that enable meaningful coexistence with these autonomous agents, ensuring that convenience does not come at the cost of human autonomy, social connection, or ethical accountability.

Living with AVs demands a recalibration of our expectations and responsibilities. The driver's seat, once a locus of control and vigilance, becomes a theatre of delegation and observation. This shift alters not only practical dynamics—how journeys unfold—but also psychological and social dimensions. Users must negotiate trust with machines whose decision-making processes remain largely inscrutable, cultivating an awareness of both the capabilities and limits of automation. The silent chauffeur's quiet competence invites both relief and unease, requiring mental frameworks that balance confidence with critical engagement.

Central to this navigation is the development of *informed trust.* Trust in autonomous systems cannot be unconditional; it must be built on transparency, reliability, and accountability. AV users benefit from understanding how algorithms interpret their environment, make choices, and handle uncertainty. This implies a need for education and clear communication from manufacturers and regulators, but also a personal willingness to engage with technological complexity rather than passively cede control. Recognising that automation is not infallible compels users to remain alert to system feedback and ready to intervene or adapt as situations evolve.

Civic strategies for integrating AVs hinge on collective governance and participatory design. Autonomous mobility is not a purely individual experience but a shared public phenomenon that intersects with urban planning, social equity, and environmental sustainability. Communities must advocate for policies that safeguard public interests, ensure equitable access, and prevent the monopolisation of mobility by private interests. For example, the allocation of AV routes and parking must reflect diverse needs rather than privileging affluent neighbourhoods. Equally, data privacy concerns demand robust protections against surveillance and commercial exploitation embedded in AV infrastructure.

The silent chauffeur also invites reflection on the *social fabric* of movement. Driving has historically been a site of human interaction—whether through gestures, signals, or shared experiences on the road. As control shifts to algorithms, opportunities for spontaneous encounters or communal rhythms risk diminution. To live meaningfully with AVs requires cultivating spaces and practices that preserve sociality in transit, perhaps through the design of communal mobility hubs, shared rides, or digital platforms that encourage human connection alongside automated navigation.

Adapting to AVs also involves confronting *ethical and psychological complexities*. Users may experience delegation fatigue, reduced situational awareness, or overtrust, all of which compromise safety and agency. Mental models that foreground active participation, such as viewing AVs as collaborators rather than mere tools, can mitigate these effects. Training programs and user interfaces should support this partnership, encouraging users to remain cognitively engaged while benefiting from automation's efficiencies.

Real-world case studies offer instructive insights. In Singapore, for instance, a phased approach to AV deployment combines pilot zones with public education campaigns and feedback mechanisms, fostering an ecosystem where citizens gradually internalise new roles alongside autonomous technology. Conversely, trials in some Western cities have exposed tensions where regulatory frameworks lag behind technology, resulting in public mistrust and fragmented adoption. These examples highlight the necessity of synchronized technological, regulatory, and social adaptation.

Beyond practical strategies lies a philosophical dimension: living with the silent chauffeur challenges us to reconsider notions of freedom, control, and responsibility in the automated age. It invites an ethic of coexistence—recognising that autonomy need not mean sole control, but rather a balanced relationship with technology that respects human dignity and societal values. This ethic demands vigilance against complacency and a commitment to shaping AV integration with intentionality and foresight.

In sum, the journey with autonomous vehicles is not solely a technological trajectory but a humanistic venture. Constructive mental models and civic strategies transform silent chauffeurs from opaque, inscrutable entities into partners in mobility—agents that serve human needs without supplanting essential human qualities. This transformation requires collective effort: from designers embedding ethical foresight, to policymakers crafting inclusive frameworks, to users cultivating mindful trust and active engagement. Only through such comprehensive adaptation can the promise of autonomous mobility be realised as a genuine enhancement of human life rather than a passive capitulation to automation.

About the Author

Rowan Hale writes at the intersection of technology, ethics, and human systems. With a background in behavioral analysis and urban systems, Hale explores how emerging technologies quietly reshape the architecture of daily life—often without our consent or awareness. Known for dissecting complex innovations with clarity and curiosity, Hale brings a deep interest in the unseen trade-offs behind convenience, automation, and digital infrastructure. The Silent Chauffeur continues Hale's mission to help readers see the world not just as it is, but as it is being subtly redesigned.

www.ingramcontent.com/pod-product-compliance
Lightning Source LLC
Chambersburg PA
CBHW031350230926
51848CB00002B/460
9789371776684